The Ladbroke Grove Rail Inquiry

Part 2 Report

The Rt Hon Lord Cullen PC

HSE BOOKS

Contents

Chapters

1	Executive summary	3
2	The Inquiry	11
3	The rail industry and its regulation	19
4	The implications of privatisation	39
5	The management and culture of safety	59
6	Railway Group Standards	79
7	Safety cases, accreditation and licensing	85
8	Railtrack and Railway Safety	109
9	The safety regulator	123
10	A rail industry safety body	155
11	An accident investigation body	161
12	Summary of recommendations	169

Appendices

1	Parties and their representatives	181
2	Witnesses	183
3	Principal documents	185
4	The relevant accidents	189
5	The models proposed by parties to the Inquiry	195
6	Safety regulation and accident investigation in aviation	203
7	Joint statement of experts on risk management	207
8	Abbreviations	215
	Inquiry team	217

Chapter 1
Executive summary

1.1   As I explain in **Chapter 2,** this report relates to Part 2 of the Inquiry which was concerned, in regard to the railways, with the management of safety and the regulatory regime.

1.2   The chapter outlines my approach to Part 2, the preparations for the Inquiry, and the procedures which were followed. It concludes with some observations about the scope for recommendations.

1.3   In **Chapter 3** I provide, as a background to the following chapters, a general overview of the organisation of the rail industry and its regulation. The chapter includes an outline of the arrangements which govern the relationship between the different members of the industry, the functions and duties of the Rail Regulator and the Strategic Rail Authority (SRA), legislation for the regulation of safety and the responsibilities of the safety regulator.

1.4   In **Chapter 4** I explore the safety implications of the disaggregation of the rail industry which was brought about by privatisation. While there has been a gradual increase in overall safety levels, there is a perception that there has been a decrease in safety. Within the workforce there is a perception that emphasis on performance has affected attitudes to safety. Safety consultants have found a pre-eminent culture of focus on train performance in terms of delays. The disparity in sanctions between those for failures in performance and those for failures in safety may well have conveyed to the industry that performance was of top priority.

1.5   The Inquiry heard evidence that fragmentation of the industry has engendered defensive or insular attitudes which hinder the identification of the underlying causes of accidents and the learning of lessons from them. Within the industry differences of culture and ways of working have developed, skills and experience have tended to become compartmentalised, the breadth of training has suffered and there has been a shortage of properly trained and competent personnel.

1.6   The Inquiry heard evidence about two areas which lie beyond the ability of any one member of the industry to deal with, namely:

         (i)   the use of system authorities for large scale projects; and

         (ii)   research and development, especially in regard to matters of strategic importance.

I discuss the problems and possible ways forward.

1.7   The evidence in regard to the use of contractors, most notably by Railtrack, was a source of considerable concern. I find, first, that the current process for the award of contracts was not being operated with due regard to the amount of training and preparation of the contract workforce. Secondly, the controls in place for the management of the work of contractors and sub-contractors were inadequate. Thirdly, there is a need for an immediate and sustained improvement by the industry in the manner in which the employees of contractors and sub-contractors are controlled. Fourthly, the argument for reduction in the number of contractors is well founded. Further, it is clear that contractors should work to exactly the same safety standards as those directly employed. Competence is of vital importance.

1.8   The chapter concludes with a discussion of the role of trade unions. I emphasise that it is the responsibility of management to ensure that the elected representatives of employees, whether they are union officials or not, have a significant role in the management of safety.

1.9   **Chapter 5** is concerned with a consideration of essential elements for the management of safety on Britain's railways. The evidence indicated that a high proportion of accidents, incidents and near misses followed unsafe actions resulting from underlying deficiencies in the management of safety.

1.10   A key factor in the industry is the prevailing culture, of which safety culture is an integral part. There is a clear link between good safety and good business.

1.11   Recognising that the first priority for a successful safety culture is leadership, I find that the fragmentation of the rail industry has made it difficult to provide leadership to the industry and for it to take united action on safety, although there are signs of improvement. I identify the need for an industry body which, with the support of the members of the industry, can take the leading role in the promotion of safety across the industry.

1.12   As regards leadership within individual companies, the evidence made it plain that it is essential that the safety commitment of senior management should be continuously visible at the working level. Much can be achieved by management undertaking regular walkabout visits. Every company should have a strategic safety management leadership team, which is led from the top and devoted to health and safety issues.

1.13   A key task for leadership is the communication of safety goals and objectives. However, if communication is to be an effective instrument in the management of safety, it has to be a two-way process, involving the workforce and giving them the sense that they are able to make a worthwhile contribution. It is not clear from the evidence to what extent safety policies play an active part in influencing safety performance. There was evidence of lack of clarity and effectiveness in rules and long-standing practice, and a variability in the effectiveness of safety meetings. A confidential reporting system such as the Confidential Incident Reporting and Analysis System (CIRAS) yields useful information, but the fact that such a system has been found to be necessary is eloquent of the lack of open communication within railway organisations.

1.14   Conflicting views were expressed about the state of morale in the rail industry, but there was general agreement that it can and must improve. Improvement in the culture of safety should bring a rise in morale. Initiatives such as reward and recognition programmes can play a significant part.

1.15   The evidence clearly demonstrated that the rail industry needs to develop its ability to behave as a learning organisation. I identify a number of areas of importance. First, identifying unsafe acts and conditions and taking prompt steps to deal with them. Secondly, applying and disseminating the lessons of accidents and incidents (including near misses). Here the evidence showed that the process was inhibited by the "blame culture", and the lack of a co-ordinated system for the collation of recommendations and ensuring that they were followed up. Thirdly, using risk assessment in order to drive improvements in safety. Fourthly, gaining benefit from the process of auditing. This has been less than fully effective. Fifthly, using data and analytical tools. The evidence showed there were weaknesses in the industry's use of these materials. Sixthly, training, with particular reference to refresher courses, into which greater effort requires to be put.

1.16   Finally, I direct attention to the desirability of the industry developing a culture in which there is a progressive movement from a situation of dependency, where management make the rules and tell employees what to do, to a situation where individuals can contribute ideas and effort, while complying with the rules and procedures, through to a position where there is a committed, dedicated team approach, with a high degree of interdependency between teams and across company boundaries.

1.17   In **Chapter 6** I set out the way in which Railway Group Standards are used and developed for system safety and safe interworking, along with non-safety matters which are of concern to the rail industry. They represent a key element in the control of risk, without exhausting what has to be done in order to meet the requirements imposed by health and safety legislation.

1.18   **Chapter 7** is primarily concerned with safety cases. I am satisfied that there is a need for the framework provided by the Safety Case Regulations, within which the duty holder demonstrates, and by reference to which it operates, its arrangements and procedures for the management of safety in a consistent and effective manner. I discuss what should be done in order to clarify the responsibilities for the control of risk at sites shared by different railway operators and at the interfaces between them.

1.19   I set out an outline of the evidence before the Inquiry about the content, preparation and uses of safety cases. While it is clear that the safety case can become over-bureaucratic, it has the potential to be a valuable tool, by, for example, bringing about a systematic approach to safety and providing a record of management's commitments to safety. The evidence showed that it <u>can</u> be a "living document", part of the direct management of safety. The discipline of producing a safety case has an important value in itself, in particular if it involves not only line management but also all other levels of the organisation. The evidence demonstrated the significance of ensuring employees' understanding and knowledge of its substance.

1.20   In regard to the auditing of safety cases, I state that I attach particular importance to what is carried out internally, where "bottom up" as well as "top down" scrutiny is extremely important.

1.21   The extension of the safety case regime so as to apply to contractors, rolling stock companies (ROSCOs) and manufacturers and suppliers was advocated by a number of parties and witnesses. However, I am not persuaded that the case for such a shift is made out, mainly in view of the risk of duplication and confusion. On the other hand I recommend in principle a system for the accreditation of companies which supply products and services for use on, or in regard to, the railways, at least where they are safety-critical.

1.22 The licensing of individuals engaged in work on the railways was also advocated. I recommend that there should be a central system of licensing of drivers and signalmen with respect to their knowledge of the rules and regulations, and, in regard to drivers, the traction for which they have been assessed as competent.

1.23 **Chapter 8** is concerned with Railtrack and Railway Safety. I trace the history of events leading up to the present position with the Safety Case Regulations 2000 and, following the modification of Railtrack's network licence, the formation of Railway Safety.

1.24 Prior to these changes Railtrack were plainly accorded a dominant position. The evidence and arguments which I have heard satisfy me that it was not appropriate that a commercial organisation, such as Railtrack became in 1996, should have the role which they possessed in relation to other commercial organisations, such as train operators. I therefore endorse what has already happened in respect of the transfer from Railtrack to the safety regulator of the function of the acceptance of safety cases of train operators, and the removal from the Safety and Standards Directorate (S&SD) of Railtrack of their functions in regard to safety cases and Railway Group Standards.

1.25 The question then is whether the process of reformation has gone far enough. Railtrack have retained the function of making recommendations as to the acceptance of safety cases, and the safety regulator has to give reasons if it is to differ. Railtrack maintain that it is necessary for both them and Railway Safety to remain under the control of their common parent, Railtrack Group plc.

1.26 Having considered in some detail the arguments put forward by Railtrack and the Association of Train Operating Companies (ATOC), I reach the view that the time has come for more radical alterations. What has been done so far has proved to be an unsatisfactory half-way house.

1.27 While Railtrack have an important interest as the controllers of the infrastructure, the current Safety Case Regulations indicate a continuing dominance on their part which is not justified. I conclude that, when the acceptance of a train operator's safety case is being considered, the safety regulator should not look to Railtrack for a recommendation but give Railtrack – along with any other operators it considers appropriate – the opportunity to make representations.

1.28 As regards Railway Safety, there appears to me to be an attempt to achieve two incompatible objectives – independence and control. Railtrack and Railway Safety appear to be drawn together in the exercise of that control. I conclude that the safety regulator should be wholly in charge of the assessment of a safety case. Railtrack should have the responsibility for carrying out, or procuring the carrying out, of the auditing which is currently required by the Safety Case Regulations. On the other hand the safety regulator should be in overall charge by, for example, reviewing the adequacy of Railtrack's auditing, carrying out its own audits to the extent that it considers appropriate, and dealing with instances of non-compliance whenever they arise. As regards the setting of Railway Group Standards, this function should be assumed by an independent body which should take full account of the knowledge, views and interests of the body of railway operators, including Railtrack as its principal member. I see no good reason why an independent body should not enjoy the confidence of the industry or that its relationship with Railtrack should be a scene for conflict. It should be well able to carry out the function with competence, efficiency and fairness.

1.29   **Chapter 9** is concerned with the safety regulator. I discuss its role and its relationship to the responsibility of members of the rail industry for the delivery of safety. I point out that a number of essential functions are beyond what the management of safety by an individual operator can achieve, for example the setting of Group Standards and the external auditing of compliance with them.

1.30   I consider whether the responsibility for accident investigation should continue to be discharged by the safety regulator. A number of criticisms of the present arrangements and their performance were advanced at the Inquiry. The one which I have found to be of critical importance is that it is inappropriate for the safety regulator to carry out the function of investigation since it might be necessary for the investigation to examine the decisions and activities of the safety regulator itself. Weighing this against potential disadvantages, I consider that the stronger argument is in favour of the responsibility being entrusted to an independent body set up for the purpose, a Railway Accident Investigation Branch (RAIB), which would be similar in constitution to the Air Accidents Investigation Branch (AAIB) and the Marine Accident Investigation Branch (MAIB).

1.31   A number of parties advocated that there should be a new safety regulator, which would assume a range of additional functions, including responsibility for Group Standards. I describe the models and set out the arguments advanced. Two important issues arose:

    (i)   the future responsibility for Group Standards; and

    (ii)   the choice of safety regulator.

1.32   As to the first of these issues, I note that Group Standards are essentially the industry's standards. It is clear that the knowledge, experience and expertise for their formulation reside in the professionals who work in the industry. A number of reasons lead me to the conclusion that the safety regulator should not assume responsibility for Group Standards. There is a risk that it would become too closely involved in the affairs of the industry, with the loss of what was referred to as "the spur to constant renewal and improvement". If the safety regulator is to discharge its own distinctive role properly, it has to be distanced, and be seen to be distanced, from the industry and its members. There is also the risk that such a change would lead towards a more prescriptive style of standard which would be less demanding in respect that it would represent only a minimum, indicative of the "light touch" which the present safety regulator regards as no longer appropriate.

1.33   As regards the second issue, there is no doubt that there have been significant deficiencies in the performance of the HMRI.   One factor has been a shortage of resources. Another has been a failure to adapt to changing conditions brought about by the disaggregation of the rail industry and the barriers created by commercial interests, allied to a difficult relationship with Railtrack.   On the other hand it is important to recognise that the safety regulator is not responsible for the management of safety. The HMRI can deploy expertise which is at home with the technological advances in the railways. They also have the benefit of forming part of the Health and Safety Executive (HSE), including the cross-fertilisation of ideas, the sharing of technical resources and the support of a well-developed regulatory framework. The independent stance of the HSE is well-established and well-recognised. I explain why I am not convinced that the Safety Regulation Group of the Civil Aviation Authority (CAA) provide a helpful parallel for railways.

1.34   A number of functions which, it was suggested, should be assumed by a new safety regulator, namely establishing and managing system authorities and funding and sponsoring research and development, do not appear to me to provide an appropriate fit with the role of a safety regulator.   They rather belong to the industry itself. Further, the models hardly address certain existing industry functions of Railway Safety which are of significant importance, such as the development of the annual Railway Group Safety Plan, which would still require to be administered by a separate body on behalf of the industry.

1.35   The deficiencies in the past performance of the HMRI, while significant, do not seem to me to reveal a question of principle as to whether the HSE are the appropriate body to be the safety regulator. Nor do they show that the HMRI are not capable, given adequate resources and effective leadership, of adequately discharging that function as part of the HSE. They have shown a capacity for self-appraisal, and are applying the lessons of past failures.   They are reviewing the extent to which they should recommend that a more intrusive and interventionist approach should be taken and, in the light of that, the extent to which they require additional resources and personnel with relevant experience and expertise in regard to the railways.   As regards leadership, there is a need for the HMRI to be placed under the direction of a new post, to be filled by a person of outstanding managerial ability, not necessarily with a railway background.

1.36   The chapter also considers the relationship between the safety regulator and the public.

1.37   I next consider a number of aspects of the relationship between the safety regulator and the Rail Regulator, in regard to Railway Group Standards and the advice given by the HSE in regard to the periodic review.

1.38   The chapter concludes with some remarks about the relationship between the safety regulator and the SRA, with respect to re-franchising and the responsibility for safety strategy.

1.39   The views which I reach point clearly to the conclusion that the function of setting Railway Group Standards should be assumed by a new rail industry safety body which is independent of both Railtrack and the safety regulator. In **Chapter 10** I discuss the role, functions, constitution and resourcing of such a body, which should be concerned not only with standards, but also with the function of accrediting and licensing and other functions for the promotion of safety within the industry, such as establishing and managing system authorities, funding and sponsoring research and development, monitoring and reporting on the industry's safety performance, developing the annual Railway Group Safety Plan, disseminating good practice and providing safety leadership. As I state, the establishment of such an industry body seems to provide an excellent opportunity to re-create part of what was lost as a result of the disaggregation of the industry. It would contribute to safety being as effectively managed and regulated as it would be if the industry were a single enterprise.

1.40   **Chapter 11** is concerned with the investigation of accidents and incidents under the new RAIB which I recommend, and the relationship between such investigations and those carried out by the police.

1.41   I endorse the view that the investigation of accidents and incidents of whatever nature should be brought under the overall control of the RAIB so that all cases would require to be reported to that organisation. Serious cases would be the subject of inquiry by the RAIB, whereas the less serious would be delegated to the industry to be dealt with by formal inquiry or formal investigation. However the RAIB would have the ability to call in any case for inquiry by themselves where that appeared to be appropriate. The sole objective of investigations should be the prevention of accidents and incidents. It would not be their purpose to apportion blame or liability.

1.42   The chapter also discusses the appointment of an independent chairman and panel members in formal inquiries by the industry, the attendance at inquiries by representatives of those affected by the accident to which the inquiry relates, and the exercise by the RAIB of a supervisory function in regard to the working of formal inquiries and formal investigations, the publication of the reports of investigations and the following up of recommendations.

1.43   As regards the relationship between such investigations and those carried out by the police, it is clear that in general the overriding public interest lies in the swift determination of the causes of rail accidents, the publication of the report and the implementation of any safety lessons. Practical difficulties arise from the fact that there is a limited pool of independent technical experts, a large number of whom may be under contract to a railway operator, such as Railtrack. There is a need for a protocol dealing with the release of technical information and access to technical experts. Lastly, I agree with the proposal that, in order to assist in achieving the objective of investigation, the statements made by witnesses in connection with RAIB inquiries and industry inquiries and investigations should not be disclosed to the police save by order of a judge.

1.44   **Chapter 12** contains a summary of my recommendations, together with my views as to the bodies which should be primarily responsible for their implementation and the periods within which they should be implemented.

Chapter 2
The Inquiry

The terms of reference

2.1   As a consequence of the crash at Ladbroke Grove junction on 5 October 1999 I was appointed on 8 October 1999 by the Health and Safety Commission (HSC), with the consent of the Deputy Prime Minister, to conduct a Public Inquiry under Section 14(2)(b) of the Health and Safety at Work etc Act 1974. My terms of reference were as follows:

> "1.   To inquire into, and draw lessons from, the accident near Paddington Station on 5.10.99, taking account of the findings of the HSE's investigations into immediate causes.
>
> 2.   To consider general experience derived from relevant accidents on the railway since the Hidden Inquiry, with a view to drawing conclusions about:
>
> (a)   factors which affect safety management
>
> (b)   the appropriateness of the current regulatory regime.
>
> 3.   In the light of the above, to make recommendations for improving safety on the future railway".

2.2   In his letter to me of the same date Mr Bill Callaghan, Chair of the HSC, wrote in regard to my terms of reference:

> "We have discussed a number of factors on which we should particularly value your consideration and recommendations. These include:
>
> • incidence and causes of SPADs
> • behavioural factors affecting driving standards (including staffing, competences, working practices, hours etc)
> • the operation of safety case arrangements in practice
> • safety issues arising from the new structure of the industry, use of contractors and sub-contractors, and arrangements between the parties
> • safety issues arising from the Government's policy substantially to increase use of the railway.
>
> We should not want you to feel constrained by this list or your broader terms of reference if you consider that other issues emerge which should be fully examined. Nor should we want you to feel that you have to make recommendations on matters which have in your view been adequately covered elsewhere e.g. in the findings of Professor Uff from the Southall Inquiry, in the independent assessments carried out by Sir David Davies for the Deputy Prime

Minister, or in investigations carried out by HSE for the Health and Safety Commission".

2.3   Shortly after my appointment I decided to hold this Inquiry in two parts dealing respectively with paras 1 and 2 of my terms of reference, along with, in each case, para 3. This part of my report deals with Part 2 of the Inquiry.

2.4   The crash at Ladbroke Grove happened at a time when the Public Inquiry into the crash at Southall on 19 September 1997, which was conducted by Professor John Uff QC, was proceeding. It became obvious that there were a number of issues which were of common concern to both Public Inquiries. In these circumstances the HSC, with the consent of the Deputy Prime Minister, appointed Professor Uff and myself to chair a separate Public Inquiry under Section 14(2)(b) into these issues, which were:

        (i)   Train Protection and Warning Systems;

        (ii)   the future application of Automatic Train Protection (ATP) systems; and

        (iii)   SPAD prevention measures.

Accordingly issues such as SPAD prevention were dealt with exclusively in the latter Inquiry, which was designated as "the Southall and Ladbroke Grove Joint Inquiry into Train Protection Systems etc". Professor Uff and I have presented a separate report on the evidence which we heard in the Joint Inquiry.

Preparation for the Inquiry

2.5   As I stated in Part 1 of my report, Mr Andrew Allberry was Secretary to the Inquiry. His Secretariat arranged for the accommodation and servicing of Part 2 and managed its documentation. By their continued hard work and helpfulness the members of the Secretariat assisted in ensuring that Part 2 of the Inquiry was run with efficiency and had a good working relationship with all who came into contact with it. I would like to express my particular gratitude to Mr Allberry for his dedication, support and advice throughout. My thanks are also due to Mrs Dorothy Gordon, of Edinburgh, who cheerfully carried the burden of typing the text of the report and the preliminary drafts and revisals.

2.6   Mr Robert Owen QC, Mr Neil Garnham, Mr Eric Brown and Ms Susan Chan acted as Counsel to the Inquiry in Part 2. Their roles were to assist me in the investigation, advise on matters of law and evidence, and present evidence to the Inquiry at its hearings. Mr Michael Fitzgerald acted as Solicitor to the Inquiry and Mr Myles Hothersall as Deputy Solicitor. All of them amply fulfilled the heavy demands which were made on them. My additional thanks are due to Mr Brown for preparing helpful accounts of the regulatory regime for railways and of the United Kingdom regulatory regime for aviation safety and aircraft accident investigation, and for assisting me by marshalling evidential material.

2.7   Mr Rod Sylvester-Evans was appointed internal consultant to the Inquiry on safety matters for the purpose of Part 2, and he and Mr Brown fulfilled key roles in the preparation of evidence relating to the management of safety and the regulation of safety respectively.

2.8   Professor Peter H McKie, CBE, formerly Chairman of DuPont (UK), and Mr Malcolm J Southgate, formerly Deputy Managing Director of Eurostar (UK), were assessors in Part 2. Their knowledge, advice and support were of considerable help to me in conducting Part 2 and preparing this part of my report. However, for both tasks I bear sole responsibility.

2.9   In planning the scope of Part 2 I had the benefit of the report of Professor Uff on the Inquiry into the Southall crash, the Department of the Environment, Transport and the Regions' (DETR) review "Railtrack's Safety and Standards Directorate", the HSE's report "The Management of Safety in Railtrack" and the report by Sir David Davies on ATP.

2.10   In order to focus on what was to be examined in Part 2, Counsel to the Inquiry drew up, with my approval and subject to some revision before the opening of Part 2, a list of questions under the following six subject headings:

- the regulatory authorities;
- Railtrack and the S&SD;
- the safety case regime;
- the relationship between the constituent parts of the rail industry;
- the management of safety; and
- railway accident investigation.

2.11   The question of the accidents to be selected as "relevant accidents" for the purposes of my terms of reference required early thought. As the basis for the selection the Inquiry team considered the reports of HSE investigations and formal inquiries into over 40 accidents since the crash at Clapham Junction on 12 December 1988 which was the subject of the Hidden Inquiry. Consideration was not confined to accidents covered by the Reporting of Injuries, Diseases and Dangerous Occurrences Regulations 1995, but included incidents which might raise similar issues. The following were selected as "relevant accidents":

- Newton Junction   21 July 1991
- Watford South   8 August 1996
- Bexley   4 February 1997
- Newton Abbott   6 March 1997
- Southall   19 September 1997
- Norton Junction   23 February 1998
- Ladbroke Grove   5 October 1999

Brief details of each of these accidents are given in Appendix 4.

2.12   On 30 March 2000 I held a preliminary hearing in regard to Part 2. One of the purposes of this hearing was to determine, by reference to Regulation 5 of the Health and Safety Inquiries (Procedure) Regulations 1975, who should be entitled to appear at Part 2 of the Inquiry. Appendix 1 sets out the representation. The other main purpose of the hearing was to set the main lines of procedure at and in connection with the Inquiry.

2.13 Since it was obvious from the nature of my remit and the matters which I intended to pursue that much of Part 2 would be taken up with discussion of differences of view and opinion, I invited the parties to set out, prior to the opening of Part 2, their position in regard to the questions set out in the list to which I have already referred. In due course each of the parties submitted a statement of case, with supporting witness statements, setting out its position and the supporting reasons, and identifying the documents on which it relied and the persons who could, if necessary, speak to the matters covered. I should add that prior to making their final submissions the parties were given the opportunity to submit any alterations to their statements of case in the light of the evidence which had been put before the Inquiry.

2.14 I made orders under Regulation 7 of the 1975 Regulations for the production of documents which were likely to be material evidence. Core bundles were assembled and copies were printed for the use of the Inquiry and representatives of the parties. Residual documents were kept in a library to which parties' representatives had access. Arrangements were also made for the advance circulation of the statements of case and the reports of expert witnesses. Parties' representatives were provided with these documents, statements and reports on the basis of a written undertaking that the material was provided and was to be used exclusively for the purposes of preparation for the Inquiry. A list of the principal documents which were referred to in the Inquiry, and are in the public domain, is given in Appendix 3.

2.15 In order to stimulate discussion and to generate ideas which could usefully be considered at the Inquiry itself, arrangements were made for the holding of a number of seminars in September-November 2000. These covered the following subjects:

- Public Perceptions of Rail Safety;
- the Civil Aviation Model of Regulation;
- Employee Perspectives on Rail Safety;
- the Japanese Model of Rail Safety;
- Developing an Effective Safety Culture; and
- Management of Change.

A number of those who participated in the seminars also gave evidence in the Inquiry. The seminars did not form part of the proceedings of the Inquiry but were open to members of the public. They were devised and conducted by Mr Owen and Mr Garnham. Summaries of the discussion at the seminars, which did not attribute statements to particular participants, were later prepared and made available for comment as part of the evidence before the Inquiry. I would like to express my gratitude to Counsel and those who participated in the seminars. It is plain that they were a fruitful source of topics for discussion at the Inquiry.

2.16 As I mentioned in para 2.13 of my report on Part 1 of the Inquiry, I held two meetings with the bereaved and injured, to enable them to bring to my attention any matters which they thought the Inquiry should consider for the improvement of safety. At these meetings a number of matters were raised which were of relevance to Part 2. I would like to thank both those who attended and the many people who wrote to me in order to suggest other points which could usefully be pursued.

2.17   At the preliminary hearing, and in a public advertisement thereafter, I invited members of the public to make written submissions to the Inquiry in regard to any of the subjects which had been selected. I have taken these submissions into account and am grateful for the response.

2.18   In accordance with the normal Government approach in the case of a major accident, the HSC indicated to me that they would consider any recommendation I might make as to the payment of legal costs of persons who were parties to the Inquiry. Those representing the bereaved and injured and the three Joint Rail Unions – the Associated Society of Locomotive Engineers and Firemen (ASLEF), the National Union of Rail, Maritime and Transport Workers (RMT) and the Transport Salaried Staffs' Association (TSSA) – applied to me for the payment of such costs from the public purse. In the event I made recommendations to the HSC in respect of the costs of the bereaved and injured and the Joint Rail Unions, in the latter case for part payment of costs. In both cases my recommendations were accepted by the HSC.

The proceedings at the Inquiry

2.19   Part 2 of the Inquiry was held at the Central Hall, Westminster. The hearings began on 31 October and finished on 20 December 2000. The evidence was presented in accordance with a programme which was circulated in advance to the parties. The opening and closing submissions were provided to the Inquiry in writing. The parties were given the opportunity of brief oral submissions in supplement.

2.20   Appendix 2 contains a list of the witnesses who provided evidence to Part 2 of the Inquiry. Oral evidence was given on oath or affirmation. Evidence was taken from certain witnesses in writing. In that event their statements were read out or summarised at the Inquiry.

2.21   The great majority of witnesses were examined in chief by Counsel to the Inquiry. Parties were asked to submit an advance estimate of the time which they expected to require for the questioning of each witness, and were reminded, where necessary, of the need to adhere to these estimates in order to ensure that the Inquiry could be conducted in an expeditious manner.

2.22   On 17 October 2000 the derailment of a Great North Eastern Railway (GNER) train occurred at Hatfield, causing the death of four passengers. Since this was currently the subject of a separate investigation, conducted jointly by the HSE and the British Transport Police (BTP), this Inquiry did not seek to take evidence in relation to it. However, in accordance with the list of questions which had been drawn up for Part 2, the Inquiry heard general evidence in regard to the management of the safety of work involving maintenance contractors.

2.23   The Inquiry was greatly assisted by the fact that a number of experts on risk assessment met and arrived at a joint statement on its use and application in the rail industry. They were Dr R A Cox, Consultant Engineer; Mr P D T O'Connor,

Consultant Engineer; Mr D C T Eves, Deputy Director-General of the HSE; Ms S A Brearley, Controller of Safety Strategy and Planning in the S&SD; Professor A W Evans, Professor of Transport Safety at the Centre for Transport Studies at University College, London; Mr P J Waite, Technical Director of Sector and Business Development, Entec UK; Dr S P Walker, Reader in Computational Mechanics at Imperial College, London and Mr Sylvester-Evans. Subject to one point on which Mr O'Connor dissented, the joint statement represented their joint views. Mr Sylvester-Evans chaired the meeting and gave evidence in regard to the joint statement.

2.24   Evidence was recorded by means of a computer-assisted transcription system which enabled the Inquiry and the parties to have not only a paper copy of the transcript of the proceedings, but also a means of access to the transcript through the use of the software. I would like to pay tribute to the skill and helpfulness of the members of the staff of Harry Counsell & Co, who ensured that this worked smoothly and efficiently. The text of the proceedings was also published through the Inquiry website.

2.25   The proceedings at the time of the opening and closing submissions were filmed for television.

The scope for recommendations

2.26   My recommendations are set out in Chapter 12. There are, however, a number of matters to which I would like to draw particular attention in this chapter.

2.27   My remit in Part 2 of the Inquiry was of an unusual nature and width. However, as parties recognised, it was no part of my remit to consider the question of the future ownership of Britain's railways. Nonetheless, in accordance with my remit, I have considered the implications of privatisation for the management of safety and the regulation of safety.

2.28   The Inquiry was greatly assisted by the participation of the Rail Regulator and the SRA, by appearing as parties and providing evidence. Neither expressed any reservation about the discussion of not only their duties with respect to safety but also their respective functions. However, mindful that this Inquiry was set up under the 1974 Act, I have taken the view that it would not be appropriate for me to make any recommendation touching on the functions, as opposed to the duties, of either of them, and have contented myself with making observations, where appropriate, as to the preferred way forward.

2.29   I have been aware throughout that it is the intention of Ministers that, pending consideration of this part of my report, decisions on potentially fundamental changes to the organisation of transport safety would not be taken. Accordingly I have proceeded on the basis that none of the possible options in regard to the regulation of safety which I consider in this part of my report is foreclosed.

2.30   The making of recommendations in regard to the future responsibility for the setting of Railway Group Standards and other railway standards has been complicated by the possible effect of the forthcoming European Directive on Rail Safety. On this point I would draw particular attention to my remarks in paras 9.67-74 and 10.30.

2.31   I would like to express my particular gratitude to the parties for the work which they put into outlining the possible options for the development of the safety regime and providing evidence and arguments in support of them. This was of great assistance in enabling me to concentrate on the essential differences and assess the best way forward.

Chapter 3
The rail industry and its regulation

Introduction

3.1   The purpose of this chapter is to provide a general overview of the organisation of the rail industry and its regulation as a background to the discussion in the following chapters. It does not set out to provide the history of the industry, but will refer briefly to past events where they are directly relevant to the present state. I include the economic regulation of the industry not only because it forms part of the total regime under which the privatised rail industry operates, but also because of the relationship between the economic and the safety regulators.

3.2   In writing of the "rail industry" I should be understood as referring to Railtrack and those other companies which are involved in using, maintaining or renewing Railtrack's network, and the construction and maintenance of rail vehicles used on it. Accordingly it is not concerned with other rail infrastructures, of which the largest is that operated by London Underground Ltd (LUL). The contents of this chapter are concerned with the following subjects:

- privatisation (paras 3.3-3.4);
- Railtrack (paras 3.5-3.9);
- the S&SD and Railway Safety (paras 3.10-3.13);
- contractors (paras 3.14-3.18);
- passenger train operating companies (paras 3.19-3.24);
- ROSCOs (paras 3.25-3.26);
- freight train operating companies (paras 3.27-3.28);
- manufacturers and suppliers (para 3.29);
- rail unions (para 3.30);
- train users (para 3.31);
- finance (paras 3.32-3.38);
- the Rail Regulator (paras 3.39-3.43);
- the SRA (paras 3.44-3.48);
- Railway Group Standards (paras 3.49-3.55);
- legislation for the regulation of safety (paras 3.56-3.67); and
- the HSC, the HSE and the HMRI (paras 3.68-3.72).

Privatisation

3.3   The privatisation of the rail industry followed the Government's White Paper published in July 1992 "New Opportunities for the Railways" (Cmnd 2012) and the enactment of the Railways Act 1993 (the 1993 Act). British Railways had functioned as a fully integrated railway operation, comprehending the design, construction, maintenance and operation of their network and the traction and rolling stock used on it. Under the 1993 Act British Railways were restructured so as to create an organisation of which the constituent parts could "migrate" into fully independent entities, to be transferred progressively into private ownership. It may be noted that Article 1 of Council Directive 91/440/EEC on the development of the railways of the European Community, which had the aim of facilitating the adaptation of the railways to the needs of the single market and increasing their efficiency, stated that this was to be achieved *inter alia* by separating the management of railway operations and infrastructure from the provision of railway transport services. However, while the separation of accounts was compulsory, organisational or institutional separation was optional.

3.4   Privatisation was initially brought about by the sale of a number of businesses, such as those now operated by the freight train operating companies (FOCs) and the ROSCOs and the franchising of newly created passenger train operating companies (TOCs). Privatisation through the sale of the Government's shares in Railtrack Group plc did not take place until 20 May 1996. By 1997 the whole of the functions of British Railways had been split up into about 100 businesses and sold. In the result the relationships between the constituent parts of what had been a single business came to be governed by commercial contracts. The regulatory framework established by the 1993 Act was based on the separation of three functions, namely economic regulation, franchising of passenger services and safety regulation, to be discharged by the Office of the Rail Regulator (ORR), the Office of Passenger Rail Franchising (OPRAF) and the HSE respectively.

Railtrack

3.5   Railtrack Group plc were incorporated on 28 February 1994 as a wholly owned subsidiary of the British Railways Board. On 1 April 1994 they became wholly owned by the Department of Transport in right of the Crown. On 20 May 1996 their shares were floated on the stock market. While the legislation included from the outset the possibility of the privatisation of Railtrack Group plc, it was not originally planned as an early eventuality.

3.6   Railtrack Group plc are the parent company of a number of wholly owned subsidiaries including Railtrack plc (Railtrack). Railtrack are the sole owner and operator of the national rail network in Great Britain. This includes about 20,000 miles of track and associated infrastructure such as signalling, together with some 40,000 bridges and tunnels. They also own a considerable number of stations and light maintenance depots.

3.7   Railtrack hold a network licence granted by the Secretary of State under the 1993 Act which authorises them, *inter alia*, to operate the network. The conditions of Railtrack's licence are enforced by the Rail Regulator, to whom Railtrack are accountable. Condition 7 of the licence requires Railtrack to maintain, renew and enhance the network "in a timely, efficient and economic manner in accordance with best practice" to meet the reasonable requirements of train operators and funders. This is subject to a test of reasonable practicability and to all relevant circumstances including Railtrack's overall financial framework. Railtrack provide train operating companies with access to their network under track access agreements.

3.8   Almost all of the Railtrack stations and light maintenance depots are leased to train operating companies. However, Railtrack themselves operate 14 mainline stations, in respect of which they hold station operator licences.

3.9   Railtrack have day to day operational responsibility for the running of the network. The operational side of Railtrack is Railtrack Line, who manage the commercial and operational activity which is necessary for maintaining and developing the infrastructure. They also maintain commercial control of train operating companies through the track access agreements and the monitoring of compliance with their railway safety cases (RSCs). There are a number of headquarters directorates and a number of regionally based zonal organisations covering the whole of Great Britain. Railtrack Line have their own safety department, the Assurance and Safety Directorate. In order to operate a new rail vehicle on Railtrack's network a train operator is required under a Railway Group Standard to obtain a certificate of authority to operate from the Rolling Stock Advisory Board of Railtrack Line. For this purpose the train operator must first obtain a certificate of engineering acceptance (see para 3.11) and thereafter route acceptance. For this latter purpose the operator requires to submit a route acceptance safety case to Railtrack Line to demonstrate the conditions under which the vehicle can be operated safely on defined routes. This process is managed by the Rolling Stock Advisory Board, with support from some safety review groups, in order to cover not only risks common to all vehicles but also route-specific risks, such as interference and gauging, and specific features of the vehicle or its intended operation.

The S&SD and Railway Safety

3.10   In accordance with what was formerly Condition 3 (now Condition 6) of their network licence, Railtrack established, and until 31 December 2000 maintained, within their organisation a S&SD. In terms of the condition the directorate was

    "…to have no commercial functions or responsibilities other than those relating to safety and standards".

Its director reported to the chairman of Railtrack and, through the Group Safety, Environment and Health Committee, to him as chairman of the Board of Railtrack Group plc. The Safety Advisory Board (SAB) of the S&SD advised the director on the development of strategy and policy, and scrutinised the directorate's operation and conduct. The S&SD provided its chairman, and its members comprised representatives of various branches of the rail industry, including the rail unions.

3.11   The main activities of the S&SD were:

    (i)   overseeing Railtrack's RSC. In recent years the greater part of the safety case was produced by Railtrack Line. It was then delivered to the S&SD for checking its compatibility with other safety cases, after which it was submitted to Her Majesty's Railway Inspectorate (the HMRI);

    (ii)   accepting the safety cases of train and station operators, i.e. apart from the safety cases of Railtrack as station operators;

(iii)   auditing Railtrack's compliance with their safety cases;

(iv)   auditing operators' compliance with their safety cases, which supplemented monitoring and checking by Railtrack Line;

(v)   developing Railway Group Standards through the production of the Railway Group Standards Code, which described the process for establishing, reviewing and changing such standards. The principle of the code is that in making changes or additions to Group Standards safety considerations take precedence, but where there is more than one way of achieving the same safety outcome, other considerations, including certain economic matters, have to be taken into account;

(vi)   developing the management of the process for creating and amending the Railway Group Standards;

(vii)   monitoring industry-wide performance of safety which was reported through the Safety Management Information System (SMIS), and producing regular safety performance reports which compared performance with strategic objectives;

(viii)   developing, in consultation with Railtrack and other members of the Railway Group, the annual Railway Group Safety Plan which contained strategic safety objectives; and

(ix)   managing part of the process for the acceptance of new rail vehicles. In accordance with Railway Group Standards the operator is required to obtain through the S&SD a certificate of conformance (to demonstrate that the vehicle conforms to the relevant mandatory requirements of the Group Standards) and a certificate of engineering acceptance (to confirm that all relevant mandatory requirements have been complied with). These certificates are issued by conformance certification bodies and vehicle acceptance bodies respectively. Such bodies are accredited by Railtrack Line as being competent to perform these functions in a safe manner. A vehicle owner, manufacturer or other third party can establish the generic acceptance of a vehicle by means of its obtaining a certificate of technical acceptance.

3.12   Following the Government's acceptance, as announced by the Deputy Prime Minister on 23 February 2000, of the report "Railtrack's Safety and Standards Directorate: Review of Main Functions and their Locations" (referred to as the "Rowlands Report"), Condition 3 of Railtrack's network licence was modified so as to require Railtrack to procure that Railtrack Group plc establish and maintain a wholly owned subsidiary, for the purpose of carrying on the "Independent Railway Safety Activity" (IRSA), in effect taking over the functions performed by the S&SD, subject to the alteration of certain responsibilities under the safety case regime, to which I will refer later in this chapter. The modification also provided for the membership of the Board of the IRSA. Their non-executive directors were collectively to have appropriate practical experience in relation to the functions of the activity, and extensive current experience of the management of safety. The chairman and four other non-executive directors were not to be employees or recent employees of the rail industry; and the remaining three non-executive directors were expected to be representative of the industry. This modification was stated to be an interim measure, subject to the outcome of the present Inquiry.

3.13   The modification to Railtrack's network licence was satisfied by the setting up of Railway Safety, a non-profit company limited by guarantee, under Sir David Davies, President of the Royal Academy of Engineering, as their independent chairman. Unlike the S&SD, Railway Safety are not part of Railtrack, but are a subsidiary of Railtrack Group plc. Subject to what I have stated in the last paragraph, they have taken over the functions of the S&SD. Their chairman does not report to the chairman of Railtrack. Railway Safety are to be funded through track access charges. The SAB perform the same functions in regard to Railway Safety as they did in regard to the S&SD.

Contractors

3.14   In accordance with Railtrack's policy, as set out in their safety case as infrastructure controller, virtually all the work of maintaining and renewing their infrastructure is performed by contractors. It may be noted that as from 1 April 1994 British Rail Infrastructure Services (BRIS), a subsidiary of British Rail, carried out the function of maintenance under contractual arrangements with Railtrack. In April 1995 the functions of BRIS were subdivided into general infrastructure maintenance and track renewals. Later a number of separate entities were formed. Currently there are seven infrastructure maintenance companies (IMCs) and six track renewal companies (TRCs) which work under contracts with Railtrack.

3.15   Maintenance and renewal work accounts for more than half of Railtrack's expenditure on the network.

3.16   These contractors employ about 2,000 sub-contractors. At any given time about 20,000 individuals are involved in the work. The Inquiry was informed that about 100,000 individuals were qualified to work on the network, of whom about 85,000 held certificates permitting them to work in red zones, i.e. while trains are running.

3.17   The contracts between Railtrack and their contractors are "unregulated" i.e. they are not subject to regulation by the Rail Regulator. The older form of maintenance contract (RT1A) was a "non-interference performance contract", in which the contractor undertook to meet objectives set out in the contract, but was not required to perform specified activities. This form of contract is being replaced with a new form of contract (IMC 2000) under which the preferred bidder is selected on the basis of technical presentation (as opposed to fixed price); a target cost is agreed and reviewed annually; and any under- or over-spend is shared between Railtrack and the contractor.

3.18   In addition some maintenance and renewal contractors have unregulated contracts with passenger train and freight operating companies for the maintenance of buildings, sidings, plant and machinery.

Passenger train operating companies

3.19   At the beginning of 2000 there were 26 passenger TOCs, wholly owned by a total of 11 parent companies.

3.20   Each of the TOCs, apart from Eurostar (UK), operates under a franchise granted to the parent company under the 1993 Act. The franchise agreement sets out the minimum level of service which is to be provided; the fares that may be charged for certain services; the level of subsidy to be paid to the franchisee; and the assets, rights and liabilities to be transferred to a successor franchisee to ensure continuity of services at the end of the franchise. After a franchise has been let, the franchisee requires to obtain approval for the operator's safety case. Save where exempt, the operator also requires to obtain a train operator's licence from the Rail Regulator. Franchise agreements are not regulated by the Rail Regulator.

3.21   The original franchises, which were granted by OPRAF, a non-Ministerial Government Department, between February 1996 and March 1997, were mostly for a period of seven years. Each of the franchisees took over from British Railways Board an existing company with a passenger train operating business. The process of the re-letting of franchises has begun. The Inquiry was told that the number and nature of the new franchises are likely to be slightly different from the previous ones. In general the period to which the new franchises relate is likely to be up to 15-20 years, subject to the giving of suitable commitments. New franchises will be open to review every five years.

3.22   In order to obtain access to specified parts of Railtrack's network, TOCs have entered into track access agreements with Railtrack Line under the 1993 Act. These are generally co-terminous with the franchises. Such agreements also contain provisions in regard to Railtrack's obligation to maintain and operate the network to a standard which would allow trains to be run in accordance with the terms of the agreement; performance and possession regimes, including arrangements for compensation for excess possessions (assessed by reference to the effect on the ability of the TOC to run its trains and the amount of notice given); the excess charges payable; and the relevant track access conditions. Similarly TOCs have entered into agreements for access to mainline stations and light maintenance depots owned by Railtrack. Access agreements, including the charges payable, are subject to the approval of the Rail Regulator before they are entered into. The same applies to amendments to them. The Rail Regulator cannot in general require such agreements to be amended (unless they so provide) but under the Transport Act 2000 (the 2000 Act) he can in certain circumstances require this.

3.23   It may be noted that Eurostar (UK), Heathrow Express, Hull Trains and LUL require a train operator's licence, but not a franchise as they were not derived from a former British Rail business. They all obtain access to certain parts of Railtrack's network.

3.24   ATOC are a trade association which was established to facilitate co-operation between TOCs; and to administer industry-wide inter-operator schemes. Their membership comprises each of the TOCs including Eurostar. By virtue of the terms of the franchise agreements or train operators' licences, a number of schemes are mandatory.

They cover matters such as ticketing and settlement, discount cards, telephone enquiry bureaux and staff travel.

ROSCOs

3.25   There are three ROSCOs, each owning or leasing 3,000-4,200 rail vehicles, being traction or rolling stock of a mixture of types and ages. The stock of British Rail was sold to the private sector in 1994. The supply of rail vehicles was separated from the provision of rail services in order to prevent train operators from having an unfair advantage at the end of the franchise, and so maintain fair competition for the new franchise. ROSCOs lease their stock to TOCs, subject to a standard Master Operating Lease Agreement (MOLA) which was drawn up at the time of privatisation. These leases are unregulated, but under the franchise agreements the franchisees could enter into key contracts, e.g. with ROSCOs, only if the other party had entered into a direct agreement with the Franchising Director. The leases are of various lengths, the prices being fixed before privatisation.

3.26   ROSCOs do not have any day to day operational responsibilities for the railways. However, they are involved in ensuring that new and refurbished stock meets safety requirements. They accredit and audit the suppliers of products. ROSCOs are responsible for rectifying design and major faults. In addition they secure the heavy maintenance, i.e. major overhauling, of their stock. This is sub-contracted to six specialist maintenance companies in the private sector, such as ADtranz or Alstom, and certain TOCs.

Freight train operating companies

3.27   There are two principal FOCs. Their operations are subject to the licensing, but not the franchising, regime. They obtain access to Railtrack's network by means of access agreements with Railtrack, under which they pay access charges.

3.28   FOCs own their own stock or lease it from ROSCOs or other parties. They also have long leases for the occupation and use of freight terminals, sidings, depots and other premises owned by Railtrack.

Manufacturers and suppliers

3.29   A large number of companies are engaged in the manufacture and maintenance of rolling stock, the manufacture and installation of infrastructure equipment, and the supplying of products and services. The latter include technical support companies, sometimes referred to as TESCOs, such as Interfleet Technology and W S Atkins. The Railway Industry Association (RIA) are the trade association for the railway supply industries. Their membership also includes ROSCOs, IMCs and TRCs.

Rail unions

3.30   A number of trade unions draw their membership from individuals employed in the rail industry, in particular ASLEF (whose members are drivers), the RMT (a range of staff, including those involved in track and signalling work and maintenance) and the TSSA (supervisors, clerical managers and technical staff). The trade unions are represented on the SAB and the Rail Industry Advisory Committee (RIAC). They also take part in the work of rewriting the Rule Book and in the National Safety Task Force which was set up in November 1999.

Train users

3.31   Since 1947 the interests of rail passengers have been represented by statutory bodies. Currently this is done by Rail Passengers' Committees (formerly Rail Users' Consultative Committees). Their chairmen, along with some additional independent members, form the Rail Passengers' Council (formerly the Central Rail Users' Consultative Committee). The work of these committees covers the investigation and determination of complaints, the tracking and reporting of trends in the quality of service performance, dialogue with members of the industry on current practice and future plans, and the representation of passengers' interests in dealing with national and local government, regulatory and funding agencies and other bodies. They are represented on the SAB and the RIAC.

Finance

3.32   Railtrack do not generally receive a direct subsidy from the Government by way of grant income. They are, however, indirectly dependent on public support as most of their principal customers, the TOCs, receive financial support in the form of subsidies and grants. The majority of the income of Railtrack is from their charging TOCs and FOCs for access to their network and mainline stations. These charges are determined by the Rail Regulator. They are designed to reflect the full economic cost of access. Passenger operator access charges are calculated on the basis of the Gross Revenue Requirement (which is the sum of the operating, maintenance and renewal costs, together with the allowed cost of capital) less Single Till Revenue (i.e. income other than passenger franchise income). Efficiency savings may be stipulated by the Rail Regulator. The allowed cost of capital is arrived at by multiplying the weighted average cost of capital by what is known as the regulated asset base, as determined by the Rail Regulator. The latter is a regulatory measure of the value which shareholders initially paid for their shares and initial debt vested in the company, which is rolled forward to include the value of subsequent investment in enhancements, together with a number of technical adjustments.

3.33   Railtrack's ability to make a profit depends on the extent that they can "out-perform" the assumptions made by the Rail Regulator as to their operating costs or receipts. In practice the principal means by which Railtrack fund major infrastructure improvements is through loans and retained profits. Railtrack's likely investment in the network is set out in their Network Management Statement.

3.34   The main sources of income for the TOCs are ticket sales and the public subsidies payable by the SRA under their franchises. In general, the first round franchise agreements provided for a reduction in subsidies over the life of the franchise. Any increase in track access charges is passed on to the SRA, and hence to the Government. In addition the SRA pay TOCs revenue support grants to assist them in paying such charges. About 70% of the costs of TOCs are access charges and the costs of leasing rail vehicles.

3.35   ROSCOs do not receive direct public funding. Their income is derived from their leasing of stock to TOCs. ROSCOs are responsible for securing the necessary finance for any new stock, and for commissioning heavy maintenance of their fleet of rail vehicles. Mandatory modifications are covered by the MOLA. If the costs incurred by any ROSCO on mandatory modifications within any one year are less than £20m, the TOCs pay 10% and the ROSCOs 90%. Only if the costs are in excess of £20m in any one year will the Government contribute. In those circumstances, the TOCs continue to pay 10%, but the ROSCOs contribute 30% and the Government the remaining 60%. This arrangement will expire in 2004.

3.36   While the operations of the FOCs are not subsidised, revenue support is provided through track access grants provided by the SRA to assist them in meeting Railtrack's track access charges. Grants may also be provided to assist in the provision of facilities for the haulage of freight by rail.

3.37   The general principle for investment funding is that Railtrack are responsible for the cost of work to the infrastructure, whereas the TOCs and the ROSCOs are responsible for rolling stock costs. Railtrack are entitled to seek inclusion of sums expended on investment within the regulatory asset base and thus increase their track access charges. This will in turn result in a requirement for an increase in the subsidy to the TOCs. The track access agreements also make provision for Railtrack to seek to have access charges reopened in the event of a change in the law. Railtrack have also received substantial funding from the European Union in respect of certain routes.

3.38   In July 2000 the Government published arrangements for increasing capital investment in the railways in their 10 year plan "Transport 2010". There will be a £7 billion Rail Modernisation Fund administered by the SRA, together with £4 billion of direct grant to Railtrack to fund renewal schemes. About £5 billion will be provided for the Channel Tunnel Rail Link. The public investment forms part of a total investment package over 10 years of £49 billion, comprising £38 billion for passenger services including train protection as presently envisaged; £7 billion for new rolling stock; and £4 billion for rail freight. The plan is based upon the premise that improvements will be driven by the re-franchising process and decisions on infrastructure enhancement. The plan set growth targets of 50% for passenger traffic and 80% for freight.

The Rail Regulator

3.39   The Rail Regulator is the independent economic regulator. He is responsible for the regulation of monopoly and dominant positions in the rail industry, in particular that of Railtrack. He does this by the issuing and enforcement of licences and by setting the terms of track access agreements between Railtrack and other railway operators. He determines Railtrack's financial framework and access charges paid to them by other railway operators, and is responsible for holding Railtrack accountable for their licence obligations. The Rail Regulator is not responsible for franchising. This is done by the SRA.

3.40   The duties of the Rail Regulator include:

(i)   the granting of train and station operators' licences and exemptions from licensing. An operator's licence may not be granted unless the operator has an accepted safety case. A licence can be modified by the Rail Regulator only with the consent of the licence holder or as a consequence of a reference to him by the Competition Commission. A licence can be modified by the Secretary of State under certain legislation. In addition the 2000 Act has enabled the Competition Commission to modify a licence in certain circumstances. It may be noted that Railtrack's network licence is a matter for the Secretary of State. The Rail Regulator enforces both licences granted by him and licences granted by the Secretary of State;

(ii)   the enforcement of licence conditions, including any relating to safety. The Rail Regulator enforces both licences granted by him and licences granted by the Secretary of State. He is under a duty to investigate alleged or apprehended contraventions, and may make orders for securing compliance. A final order may impose a monetary penalty. Failure to comply constitutes breach of statutory duty, and may provide a ground for revoking the licence. His decisions are subject to a right of appeal. The Rail Regulator is not required to take enforcement action if the contravention is trivial, if the public interest requires a different course of action, or if the licence holder is taking remedial action;

(iii)   the supervision of the granting of access to Railtrack's infrastructure, stations and depots, including the approval of the terms of, and the charges under, access agreements and their periodic review. His approval is essential to the validity of an access agreement. Under the track access conditions the Rail Regulator has an appellate jurisdiction in regard to the costs of the network and vehicle changes;

(iv)   the approval of the Railway Group Standards Code;

(v)   the undertaking of responsibilities under competition legislation; and

(vi)   under a new provision contained in the 2000 Act, the making of directions for the provision of a new railway facility or the improvement or development of an existing one.

3.41   The Rail Regulator's general duties include a number directed to the protection of the interests of the users of railway services and the promotion of the use, and efficiency, economy and competition in the provision, of such services. He is also under a duty to have regard to the financial position of the SRA. To these duties the 2000 Act has added duties to facilitate the strategies of the SRA; to contribute to the development of an integrated system of transport of passengers and goods and the achievement of sustainable development; and to have regard to general guidance by the Secretary of State about railway services or other matters relating to railways.

3.42 While the Rail Regulator is not a funder, he has a duty to ensure that Railtrack are properly funded as a competent and efficient organisation. He is responsible for ensuring that Railtrack maintain and renew the network properly, and that the regulatory regime provides adequate incentives to ensure that they do. In his periodic review of Railtrack's access charges which was published on 23 October 2000 the Rail Regulator rewrote the financial framework for investment in the railways and set the charges which Railtrack are permitted to make in the five years starting in April 2001. He also replaced the existing contractual regime for bonuses and compensation with a new performance regime.

3.43 Under Section 4(3)(a) of the 1993 Act the Rail Regulator has:

> "…to take into account the need to protect all persons from dangers arising from the operation of railways, taking into account, in particular, any advice given to him in that behalf by the Health and Safety Executive".

Section 151(7) of the 1993 Act provides that nothing in the Act or done under it is to prejudice or affect the operation of any of "the relevant statutory provisions" as defined in Part 1 of the Health and Safety at Work etc Act 1974 (the 1974 Act). The relationship between the Rail Regulator and the HSE is governed by a memorandum of understanding.

The SRA

3.44 The establishment of the SRA was put forward in the Government White Paper "A New Deal for Transport: Better for Everyone" in 1998. They existed in shadow form prior to coming into existence on 1 February 2001. Section 202 of the 2000 Act constituted the SRA as a non-departmental body with a board of 15 members with, under Section 205, the following purposes:

> "(a) to promote the use of the railway network for the carriage of passengers and goods,
>
> (b) to secure the development of the railway network, and
>
> (c) to contribute to the development of an integrated system of transport of passengers and goods".

Under the Act the SRA are authorised to make investments and loans, give guarantees and provide grants. They have also assumed the role previously discharged by the Franchising Director; the residual functions of the British Railways Board; responsibility for consumer protection and the sponsorship of the network of rail users bodies, previously exercised by the Rail Regulator; and the function of providing grants to the freight industry which was previously administered by the DETR. The

Chairman of the SRA, Sir Alastair Morton, described the SRA in evidence as "the instrument by which Government policy in regard to the railways is effected". This is to be achieved by a control of franchise agreements and the provision of funding.

3.45   The SRA are to formulate, and keep under review, strategies with respect to their purposes. They are to act in the way which is best calculated to advance certain purposes, including the protection of the interests of users and the promotion of efficiency and economy in the provision of railway services. They are subject to directions and guidance from the Secretary of State in respect of their strategies and the exercise of their functions. They are further required to secure that any grants, loans or other payments made by them are such that they reasonably consider will further their purposes efficiently and economically.

3.46   Under Section 207(3)(a) of the 2000 Act the SRA are to have regard to, *inter alia*,

> "...the need to protect all persons from dangers arising from the operation of railways (including, in particular, by taking into account any advice given by the Health and Safety Executive)".

3.47   The Inquiry was informed by the SRA that their policy is to use franchising agreements

> "...as a means of driving up safety standards and enforcing safety compliance by franchise operators".

Prospective franchisees will be asked to propose safety commitments. Franchise agreements will require franchisees to implement specific commitments and to use all reasonable endeavours to improve the operator's safety record and the safety standards of the franchised services on a continual basis, and to submit an annual safety plan identifying specific targets for improvement. The SRA will monitor safety performance. Failure to meet commitments may result in the revocation of the franchise. The SRA will have a Safety and Research Panel, its remit being:

> "...to consider all issues relevant to the enhancement of strategic railway safety and the conduct of strategic research in the railway industry and to recommend issues for policy consideration to the main Board".

The panel will be advised by the Safety Director (who is part of the Infrastructure Directorate of the SRA) on anything which is found in annual safety reviews of the franchisees.

3.48   It is anticipated that an existing memorandum of understanding between the Franchising Director and the HSE in regard to liaison and co-operation on safety issues will be superseded by an updated memorandum which reflects the new role of the SRA and changes in the safety case regime. It will include a mechanism by which the HSE are to be involved in franchising decisions. In the event of a conflict of views on the matter of safety, that of the HSE is to take precedence over the SRA's. Franchise bids are already sent to the HSE (and Railway Safety) for their comment; and both bodies are involved at a number of stages in the franchising process.

Railway Group Standards

3.49   The system for the management of safety within the rail industry is dependent on the setting of, and compliance with, Railway Group Standards, and safety assurance based on auditing, together with the investigation of accidents and incidents and the taking of any necessary remedial action.

3.50   Railway Group Standards have their origin in the rules developed by British Railways for the discharge of their responsibilities. They should be distinguished from the more detailed Line Standards which are set by Railtrack Line for the purpose of achieving their own compliance with Railway Group Standards.

3.51   After their formation the S&SD overhauled the process for the development of standards, and set up the Railway Industry Standards Strategy Committee (RISSC), which is a pan-industry group, to advise on policy regarding strategic issues relating to Railway Group Standards. The S&SD received advice also from subject committees consisting of experts in a particular railway discipline, such as the Traction and Rolling Stock subject committee. As I have already narrated, the S&SD also developed the Railway Group Standards Code. Under this Code, a Group Standard can be imposed only if, on the basis of a cost benefit analysis, its safety benefits exceed the cost of its implementation. The setting and enforcement of a standard can impose a heavier burden on some members of the Railway Group than on others. Section 12 of the Code contains an appeal procedure in respect of changes to Group Standards. Railtrack's network licence has recently been modified by the Rail Regulator to the effect of extending the right of appeal against new or amended Railway Group Standards to "any stakeholder", i.e. any person having a safety case; any funder; and any person whose business activities or goods which he manufacturers must comply with Railway Group Standards. The initial stage is an appeal to a senior manager in the S&SD not involved in the detail of the setting of the standard. A further appeal lies to the Rail Regulator. However, no instance has yet occurred of this procedure being used.

3.52   The S&SD undertook the reducing of 9,000 or so standards which had been inherited from British Railways to a much smaller number of standards which are generally goal-setting and less detailed than their predecessors. According to the evidence given on 27 November 2000 by Mr R I Muttram, Director of the S&SD, the number had been reduced by them to just over 500, with a number still under revision.

3.53   Railway Group Standards set out requirements for system safety and safe interworking. They consist of:

      (i)   technical standards with which railway assets or equipment used on or as part of railway assets must conform; and

      (ii)   operating and management procedures with which all operators of railway assets, including Railtrack, must comply.

The standards deal with the control of risks to passengers, railway workers and members of the public arising from railway specific hazards which are associated with infrastructure operations and train movements; and station operations, to the extent that they affect safe train operations or the movement of passengers to and from trains. They include the Rule Book, which is in the course of being re-written. Nearly all the standards are concerned, in whole or in part, with safety. However, certain aspects of the standards relate to matters which are not safety related. These include processes, procedures, information obligations and provisions with respect to network harmonisation, all of which are primarily matters of economic regulation.

3.54 Railway Group Standards are mandatory for members of the Railway Group, although the Safety Case Regulations do not require the duty holder to comply with them. However, in the case of the TOCs, compliance with them is regarded as mandatory under their track access agreements with Railtrack, as well as their operating licences. In the case of Railtrack, it is a condition of their network licence. As from 1 October 1996 membership of the Railway Group was confined, as a result of a decision of Railtrack, to companies holding an accepted safety case. It may be noted that Group Standards also apply to the FOCs, Eurostar (UK), Heathrow Express, Hull Trains and LUL. Group Standards are enforceable directly by the Rail Regulator, by reason of their being a condition of the licence granted by him. They may also be enforceable indirectly by the HSE as the safety regulator where compliance with them is undertaken under the safety case. They are not enforceable by either the Rail Regulator or the HSE against other companies which do not have an operator's licence or a RSC. The general import of the memorandum of understanding between the Rail Regulator and the HSE is that the latter would normally be ready to take action in respect of the safety element of a Group Standard, whereas, to the extent that it is a matter of a purely economic nature, the Rail Regulator would expect to take responsibility for its enforcement. Such standards may also be imposed by a member of the Railway Group as a condition of contract with a non-member, for example by Railtrack in their contracts with contractors working on the infrastructure.

3.55 It is important to note, as I pointed out in para 14.4 of my report on Part 1 of this Inquiry, that Railway Group Standards do not apply to matters which are entirely within the control of a TOC and do not affect the safety of the staff or passengers of any other TOC or the general public. Thus a number of standards which had been issued by British Railways were withdrawn some years later and thereafter treated as placed in the public domain for access by all members of the rail industry. However, ATOC have taken up these matters with a view to issuing their own standards, supported by approved codes of practice and guidance. Mr Muttram gave evidence in Part 2 of this Inquiry that, with the assistance of the S&SD, a number of standards had been recovered from the common domain. These would probably be reduced to about 30 or 40 key standards in due course.

Legislation for the regulation of safety

3.56 The activities of companies in the rail industry are subject to the general law relating to health and safety. Sections 2-4 of the 1974 Act imposed general duties on employers to their employees, and on employers and persons concerned with "premises" to persons other than their employees. The application of the Act to transport systems and the public was clarified by Sub-section (2) of Section 117 of the 1993 Act which applied to, *inter alia*, any railway. It stated that the general purposes of the 1974 Act included:

"(a)   securing the proper construction and safe operation of transport systems to which this section applies, and of any locomotives, rolling stock or other vehicles used, or to be used, on those systems; and

(b)   protecting the public (whether passengers or not) from personal injury and the risks arising from the construction and operation of transport systems to which this section applies".

Sub-section (1) of the same section also made railway legislation, as set out in Sub-section (4), "existing statutory provisions" for the purposes of the 1974 Act, and accordingly subject to repeal or modification by regulations made under that Act. Other duties imposed under health and safety legislation which are of general application include, for example, the Management of Health and Safety at Work Regulations 1999 (which replaced earlier regulations of 1992) and the Reporting of Injuries, Diseases and Dangerous Occurrences Regulations 1995 (which replaced earlier regulations of 1985).

3.57   The application of general legislation is supported by regulations which are specific to railways. The most fundamental of these regulations are the Railways (Safety Case) Regulations 2000, which replaced the earlier regulations of 1994 as from 31 December 2000.

3.58   Other regulations relating to railways include the following:

(i)   The Railways (Safety Critical Work) Regulations 1994, which apply to the work of drivers, guards, conductors, signalmen and certain other work in a maintenance or supervisory capacity. The regulations require that individuals undertaking such work are competent and fit to do so, and that arrangements are in place to enable railway operators and the HSE to check records of this. They also enable the HSE to approve assessments of fitness and competence; but this power has not been exercised;

(ii)   The Railways and Other Transport Systems (Approval of Works, Plant and Equipment) Regulations 1994, which require the approval of the Secretary of State (in practice the Chief Inspecting Officer of the HMRI and his deputy) before new works, plant or equipment, or any alteration to any existing works, plant or equipment, may be brought into use for the purposes of a relevant transport system, including a railway. As I noted in my report on Part 1 of this Inquiry at para 10.4, the requirement for advance approval is qualified by the terms of Regulation 4(4)(a). The requirement for the approval of any new rail vehicle includes its interior and is in addition to the requirement imposed by Railtrack, to which I have already referred. It should be noted that the Rail Vehicle Accessibility Regulations 1998 impose construction requirements for rail vehicles first brought into use from 1 January 1999, and impose certain duties on operators to assist disabled passengers. The purpose of these regulations is to allow disabled people, including wheelchair users who remain in their wheelchairs, to get on and off rail vehicles in safety and without unreasonable difficulty and to travel in them in safety and reasonable comfort. The Department of Transport, Local Government and the Regions (DTLR) advise operators and manufacturers in regard to compliance, and are responsible for an exemption process;

(iii) The Railway Safety (Miscellaneous Provisions) Regulations 1997, in addition to repealing certain railway legislation as outdated, require infrastructure controllers to ensure, so far as is reasonably practicable, that procedures are in place and equipment is provided and maintained for the purpose of preventing collisions and derailments; and require vehicle operators to provide and maintain suitable and sufficient braking systems; and

(iv) The Railway Safety Regulations 1999 contain provisions in regard to the use of train protection systems, Mark I rolling stock and rolling stock with hinged doors.

3.59 Following the Government's acceptance of the Rowlands Report, steps were taken to alter the safety case regime, resulting in the replacement of the 1994 with the 2000 Regulations in regard to safety cases. This change was also expressed to be an interim measure pending the outcome of the present Inquiry. The main features of the system under the 2000 Regulations are as follows.

3.60 The regulations require that infrastructure controllers, train operators, and station operators should have a valid safety case. The general object of a safety case is to ensure that an operator has the will, capabilities, organisation, system and resources to operate safely. Schedules 1 and 2 to the Regulations make provision as to the particulars which are to be included in a safety case. In this respect certain additional requirements have been introduced by the 2000 Regulations. The holder of a safety case has a duty to conform with it. Failure to do so is a criminal offence. The category of train operators includes contractors to the extent that they operate steel-wheeled vehicles on the railways, whether in or out of possession. Subject to that qualification, IMCs and TRCs are not required by the regulations to have a safety case. However, Railtrack require them as a condition of contract to submit a contractor's safety case, which is currently being replaced by a contractor's assurance case.

3.61 It is essential to the validity of a safety case that it is accepted by an external body. Without such acceptance the duty holder cannot lawfully operate. Under the 1994 Regulations the infrastructure controller (in practice the S&SD in the case of Railtrack) was responsible for acceptance of the safety cases of train and most station operators. This function now requires to be discharged by the HSE. An appeal against a refusal to accept (actual or deemed) lies to the Secretary of State.

3.62 Under the 2000 Regulations the infrastructure controller has to procure an assessment of its own safety case by an assessment body (in practice Railway Safety in the case of Railtrack) and to obtain and submit to the HSE a report of that assessment, including a recommendation as to whether the safety case should or should not be accepted. Where acceptance is not recommended the reasons for that have to be stated. The safety case which is prepared by a train or station operator has to cover all the duties of the operator and not merely those which are of concern to the infrastructure controller. Accordingly it will include matters such as those relating to the interior design of carriages. However, under the 2000 Regulations the assessment of such a safety case, which the infrastructure controller has to procure along with a report on it, and its own recommendation, are restricted to a consideration of whether the procedures and arrangements described in it will, when properly implemented with those described in any other safety case or revision thereof, be capable of ensuring compliance by the infrastructure controller with its own health and safety duties in relation to the operation to which the train or station operator's safety case relates. From this and the previous paragraph it can be seen that the role of Railway Safety in regard to the acceptance of safety cases, unlike that of the S&SD, is limited to being an advisory one.

3.63   The duty holder is required to keep a copy of the accepted safety case and any revision thereof available for public inspection.

3.64   The holder of an accepted safety case has a duty to review it at least every three years: and has a duty to revise it:

  (a)   to meet the additional requirements of the 2000 Regulations;

  (b)   whenever it is appropriate; and

  (c)   whenever required to do so by the HSE.

A revision which will render the safety case materially different from the version last accepted is not to be made unless the HSE have accepted it. In the case of the safety case of a train or station operator, the infrastructure controller is to provide the HSE with its recommendation as to whether the revision should or should not be accepted and the reasons for that recommendation. An appeal against a direction to submit a revision and a refusal (actual or deemed) to accept a revision lies to the Secretary of State.

3.65   It may be noted that the infrastructure controller is to take all reasonable steps to ensure that train and station operators conform with those parts of their safety case which affect or are likely to affect the performance of their health and safety duties. It has also to notify the HSE if it is aware that an operator is failing to comply, in consequence of which the risk of serious injury is increased; or if the operator fails to comply with a reasonable request in respect of any aspect of the operation which affects or is likely to affect the performance of the health and safety duties of the infrastructure controller, contrary to an undertaking which it is required to obtain before permitting the operator to run a train or operate a station.

3.66   The 2000 Regulations also introduce a statutory duty of the infrastructure controller to procure the assessment body to undertake, at intervals of not more than 12 months, an audit of:

  (i)   the operations of the infrastructure controller; and

  (ii)   the operations of any train or station operator in relation to the infrastructure controller's infrastructure.

The audit report is to be copied to the HSE, the operator and any other operator affected by the matters to which the report relates. Copies are to be kept at a notified address. An audit is:

> "...a systematic assessment of the adequacy of the management system of the railway operator to achieve compliance by him with the relevant statutory provisions in relation to the operations undertaken by him".

3.67   Certain duties of co-operation are imposed under the 2000 Regulations. Thus for example, a contractor carrying out work on or in relation to premises or plant owned or controlled by a railway operator has a duty to co-operate so far as is necessary with the operator to enable him to comply with the Regulations.

The HSC, the HSE and the HMRI

3.68   The HSC are a statutory body consisting of a chairman and a number of members appointed by the Secretary of State. The latter are chosen after consultation with a number of representative and other organisations. Under Section 11(2) of the 1974 Act they have a number of general functions, including making arrangements for research and the provision of training and information in connection with the general purposes of the Act, the provision of information and advice to the public and private sectors, and the submission of proposals for the making of regulations. The Commission are responsible (with the assistance of the HSE) for the formulation and dissemination of policy for the regulation of safety. The Government require to consult the HSC on health and safety matters in connection with legislative proposals, and the HSC are also required to consult with key domestic stakeholders before tendering advice to Ministers. The HSC also have the benefit of advice from the RIAC. The membership of RIAC is drawn from a wide range of interests in the rail industry, along with representatives of the three main trade unions and passengers' interests.

3.69   The HSE are in part the "operating arm" of the HSC. Under Section 11(4) of the 1974 Act they have the duty to exercise whatever functions the HSC direct them to exercise on their behalf, and to give effect to any directions which the HSC have issued. On the other hand, under Section 18 of the 1974 Act the HSE have a direct statutory responsibility for the enforcement of the relevant statutory provisions.

3.70   The HMRI acquired their current title on 3 December 1990 when they were transferred from the Department of Transport to the HSE and became the HSE's operational division with responsibility for health and safety on the railways. Under an agency agreement between the HSC and the Secretary of State which came into force at that time, the HSE undertook through the HMRI the regulation of certain functions on behalf of the HSC and the Secretary of State. In May 2000 the HMRI were "brigaded" within the Field Operations Directorate of the HSE. The HMRI publish an annual report on the safety record of Britain' railways, the work of the HMRI and areas of concern. The HMRI provide an extensive advisory service to the rail industry, ranging from written guidance to the answering of telephone calls. They have promulgated high level principles and guidance in "Railway Safety Principles and Guidance", which was published in 1996. As from 6 December 2000 the HSE brought into effect new arrangements comprising the formation of a Key Railways Issues Group for the co-ordination of all railway actions of the HSE, and a Railways Directorate for working with operational inspectors to promote and regulate rail safety.

3.71   While it is the case that responsibility for the regulation of safety is laid upon the HSC acting along with the HSE, it is convenient for the sake of brevity to refer to the HSE as the safety regulator, their responsibility being discharged largely through the HMRI.

3.72   It may be useful at this point to set out in brief the position in regard to a number of functions which will be considered in more detail in Chapter 9:

(i)   under and by virtue of Section 18 of the 1974 Act the HSE have a direct statutory responsibility for the enforcement of legislation affecting railways. Their inspectors have the power to inspect and monitor the safety compliance of members of the rail industry and to issue improvement and prohibition notices (Sections 20-22). The HSE may initiate, and in England, but not in Scotland, conduct, prosecutions for breaches of health and safety legislation, including legislation specifically relating to railways. However, it should be noted that, unlike the Rail Regulator, the HSE have no power themselves to impose financial penalties;

(ii)   as I have already stated, the HSE are responsible for the approval of all safety cases, along with their revisions, in accordance with the 2000 Regulations;

(iii)   the HSE are responsible for the granting of approvals under the Railways and Other Transport Systems (Approval of Works, Plant and Equipment) Regulations 1994; and

(iv)   the statutory responsibility for making arrangements for the investigation of, or inquiries into, any accident, occurrence, situation or any other matter which is thought to be necessary or expedient to investigate lies with the HSC under Section 14 of the 1974 Act. The HSC may direct the HSE or authorise any other person to investigate and make a special report or, with the consent of the Secretary of State, to conduct an inquiry, such as the present Inquiry.

Chapter 4
The implications of privatisation

Introduction

4.1   In the light of my terms of reference I require to consider the implications of privatisation, and whether it has led, directly or indirectly, to a deterioration in the safety performance of the industry.

4.2   The issues which I consider in this chapter are as follows:

- the background (paras 4.3-4.11);
- commercial considerations and performance (paras 4.12-4.19);
- the effects of the fragmentation of the industry (paras 4.20-4.54);
- the franchising process (paras 4.55-4.60);
- the use of contractors (paras 4.61-4.88); and
- the role of the trade unions (paras 4.89-4.101).

The background

4.3   The process of privatisation was initiated in July 1992, when Ministers published a White Paper setting out their proposals for legislation. The White Paper was followed in January 1993 by the report by the HSC "Ensuring Safety on Britain's Railways". This paper developed proposals for assuring safety following the liberalisation of access to and the privatisation of British Railways. Paras 5 and 6 of the executive summary of this report stated:

> "Control of railway operations will be divided between many different organisations and this will generate a need to define the extent of the responsibilities of each party and to ensure effective management of safety (including emergency planning) particularly at the numerous interfaces between parties.
>
> Unless considerable care is taken to set up systems to ensure that new operators are properly equipped and organised there can be no confidence that risk will be effectively controlled right from the start and that important matters do not fall between the safety arrangements of the various parties. The consequences of failing to achieve adequate systems of control will be seen in increased risk on the railway system and the likelihood of an increase in the numbers, and possibly also the severity, of accidents".

4.4   The Inquiry was provided with clear evidence on overall safety trends since privatisation. The statistics do not bear out a picture of a declining safety trend. Professor A W Evans, Professor of Transport Safety at the Centre for Transport Studies at University College, London, had made an extensive statistical analysis of the safety performance of the railways before and after privatisation. He concluded that safety performance was the same before and after privatisation. He stated: "I would not say that safety has deteriorated". Sir David Davies, President of the Royal Academy of Engineering, supported this view in his report on ATP for the Railway Network in Britain, published in February 2000. He stated:

> "There has been a continuing (but gradual) increase of safety levels over a period of three decades".

4.5   The Joint Rail Unions, in their statement of case to the Inquiry, quoted a report on worker safety in the rail industry by Professor C J Baldry, Professor of Human Resource Management at the University of Stirling, and Dr J Ellison of the Department of Management and Organisation of the University of Stirling. This report stated:

> "The falling trend in fatalities to both passengers and staff and the decline in major injuries in the industry, despite the disasters of the last three years, still compare favourably with the earlier BR years pre-Clapham".

4.6   There was no evidence from which I could conclude that, whatever the way in which privatisation was carried into effect, it would be detrimental to safety. However, some parties and witnesses maintained that the way privatisation had been carried out had had a wide range of consequences that were, directly or indirectly, detrimental to safety. This latter view was mirrored by clear concerns which were voiced at the Inquiry's seminar on Public Perceptions of Rail Safety. A number of those attending this seminar felt that a lower priority was given to safety than to other business parameters. The public perception, reinforced by the crashes at Southall and Ladbroke Grove, seemed to be clearly that the safety trend was downwards. At the seminar on Employee Perspectives on Rail Safety trade union representatives spoke of their members' perception of pressures in regard to performance and how this affected attitudes to safety.

4.7   In Part 1 of this Inquiry there was evidence that traffic volumes had increased by 20% per year since privatisation, and that this trend was expected to continue. The parties were requested to comment on the safety implications of this increase in traffic.

4.8   Amey Rail, the HSE, the Rail Users' Committee, Railtrack and the Joint Rail Unions all drew attention to the expected need for more maintenance to deal with increased wear and tear, and pointed out that there would be less time available to execute this maintenance. Specifically the unions were concerned about the pressure which the increase in traffic would put on the ability of Railtrack to grant possessions of the track for maintenance.

4.9   The ROSCOs, the Collins Passengers' Group and Railtrack stressed the demands which were made by the need to operate at higher levels of efficiency. English, Welsh and Scottish Railway (EWS) and Railtrack pointed out that with the increase in movements went an increase in the potential for error. The HSE also pointed out that as timetables were compressed there would be less distance between trains.

4.10   ATOC considered that all of the above concerns could be managed, but they stressed the need for "a cohesive network policy".

4.11 The fact that considerable investment would be required to achieve this increase in traffic was stressed. Counsel for the bereaved and injured represented by the Southall and Ladbroke Grove Solicitors' Group pointed out that this new investment would be expected to introduce new and improved equipment such as train protection systems with a consequential safety benefit.

Commercial considerations and performance

4.12 The Inquiry asked the parties to consider whether or not commercial considerations had had an adverse impact on safety management and whether there was a culture of "profit before safety" in the industry. It also asked if the system of penalties and compensation incorporated in the current regime of franchised operations conflicted with the proper management of safety.

4.13 The Rail Regulator, Mr T P Winsor, stated:

> "If Railtrack is a competent and efficient company, managed well, then there will be no conflict between safety and performance because they are two sides of the same coin".

In this way he considered there was no tension between performance and safety. He went on to point out that he was in no doubt that a punctual railway was a safer railway.

4.14 Nevertheless there was clearly a widely held view that commercial considerations had had an effect on safety management in the industry. While there was no specific evidence presented in Part 2 of the Inquiry that performance was in practice put before safety, the DuPont report "Safety Management in the Railway Group", which was commissioned by Railtrack early in November 1999, found that:

> "...the pre-eminent culture within the rail industry in the UK is one of focus on train performance in terms of delays",

and that:

> "...the noise about performance drowned out the noise about safety".

This view was echoed at the seminar on Employee Perspectives on Rail Safety, which heard of a clear perception at the working level in the industry that performance was given a higher priority.

4.15 Sir Alastair Morton, Chairman of the SRA, stated that in his opinion there was current tension between safety on the one hand and performance and growth on the other. He believed the tension – which was temporary, and changing – was due to a conflict in recovery priorities. At the same time he considered there did not need to be any such tension between these objectives. The important thing was to avoid undue tension in achieving the objectives.

4.16   Mr G M N Corbett, then Chief Executive of Railtrack, expressed a similar view that, in practice, this tension did exist. Giving evidence on 10 November 2000 he told the Inquiry that the demands on performance were "quite colossal". Performance, he said, was:

> "...the Government's main objective, and that is what everyone thought the passengers wanted, a big focus on train performance".

He went on to say that when he gave evidence in Part 1 on 18 July 2000, he did not believe that this strong emphasis on performance was increasing the safety risk. However, in the light of what he had heard in recent months, after the Hatfield crash, from front line workers, he said:

> "One does have to question whether the noise level at the front line actually has generated a set of behaviours which we did not know about and did not intend".

4.17   When Mr Corbett was specifically asked whether he now accepted that the attempt to adhere to performance targets might have adversely affected safety, he answered: "Yes".

4.18   Mr R I Muttram, Director of the S&SD, expressed a similar view. He said:

> "Whilst I do not believe in the long term there is a conflict between punctuality and safety, the way in which you achieve a punctual and safe railway that has been historically under-invested, does mean that care needs to be taken in the methods that one uses for moving performance forward".

He considered that the Rail Regulator should have consulted the HSE before he issued his enforcement order on punctuality targets.

4.19   When I consider these views along with, in particular, the mechanisms for the imposition of penalties for poor performance, in consequence of which fines of millions of pounds may be imposed by the Rail Regulator, and contrast these penalties, actual and potential, against the level of fines imposed in criminal proceedings in respect of serious breaches of health and safety, I conclude that the magnitude of the penalties that are likely to be imposed for poor performance, and the gross disparity which exists between performance and safety sanctions respectively, may well have conveyed to the industry the message that performance was the top priority.

The effects of the fragmentation of the industry

4.20   The Inquiry asked parties what, if any, deficiencies in the management of safety had resulted from the fragmentation of the industry. Amey Rail, ATOC and the RIA identified no deficiencies. Others, including the bereaved and injured and the Joint Rail Unions, held a strong belief that fragmentation had compromised safety, leading to a number of specific problems. Amongst more general problems such as the loss of common objectives and of a common culture and a lack of leadership, particular problems mentioned included:

- defensive or insular attitudes;
- numerous, complex interfaces, leading to problems of staff skills and experience;
- the management, development and implementation of large scale projects; and
- uncoordinated processes in research and development, with little real research and development being carried out.

I discuss these problems below.

4.21 ATOC submitted, however, that some long-standing safety problems were inherited with privatisation. They referred to a lack of cohesion in strategy and implementation and gave as an example the long-running ATP project. They also pointed to a lack of significant investment by British Rail over many years.

4.22 A mid-point view was expressed by Mr Muttram on behalf of Railtrack, who considered that the fragmentation of the industry caused difficulty in terms of safety management systems, but, on the other hand, that it had led to a greater clarity in regard to responsibility and accountability.

4.23 The fragmentation of the industry obviously presents a question as to where the overall responsibility for strategic leadership lies. I will return to this question when considering the broad aspects of safety leadership in Chapter 5.

The risk of defensive or insular attitudes

4.24 The problems which, it was claimed, were caused by the industry's loss of a common objective and attitude were illustrated in a report by Entec UK, commissioned by the Inquiry, which examined the causes of and responses to earlier accidents. Mr P J Waite, Technical Director of Entec, speaking to the report, told the Inquiry that investigations into relevant accidents often concentrated on local faults:

> "The industry-wide blame culture and fragmentation has hindered the efforts to identify the fundamental problems. The immediate cause may only be a symptom of the underlying cause".

Mr B R Burdsall, Managing Director of Midland Main Line, said that he considered that "there was a great deal of truth in that statement".

4.25 Mr K Bird, Chairman of c2c Rail, speaking to the ATOC statement of case, agreed that in the investigation of SPADs there could be a problem if the causal analysis identified a problem at the interface between, for example, the ROSCO and the infrastructure company. He also agreed that the DuPont report had identified the need for further training of driver standards managers in root cause analysis, and said that his company planned to do this. It is obviously important that incidents are fully investigated to determine their root cause and that common processes should be used across the industry to facilitate this. I will return to this subject in Chapter 5.

4.26 Regarding the follow-up of recommendations, Mr Bird agreed that there was little or no co-ordination between companies, and felt that there should be a central body to ensure that action was taken. It is my belief, however, that the follow-up of incidents should remain primarily a task for the line management of the company to which the recommendation is directed. At the same time the means must be found to ensure that the industry as a whole learns the wider lessons.

Interfaces and the number of franchises

4.27 Fragmentation, and in particular the large number of franchises operating on the railways, were said by some to have led to growing difficulties in regard to the skills and experience of staff, and cross-industry contacts between them.

4.28 The Joint Rail Unions were concerned over the growth, across franchises, of different cultures and different ways of working. These themes were echoed at the seminar on Employee Perspectives on Rail Safety, where it was argued that skills were being compartmentalised, and training and competency were becoming more limited. One example cited was that signallers did not need to have knowledge of maintenance or access to the track. It was also claimed that training to provide a general grounding in railway knowledge had stopped.

4.29 Mr S K Baker, Deputy Managing Director of Northern Spirit, predicted that the supply of managers and employees trained in the days of British Rail to have broad experience of railway operations was going to run out. The result would be more specialists and fewer generalists. He considered that Railtrack's line managers no longer understood about train working, nor train operators about the infrastructure. Mr R H McClean, Production Director of GNER, was equally concerned at compartmentalisation and the loss of the process for the generation of good quality managers with broad experience. Referring to the practice of British Rail he said that

"…it certainly generated a cadre of managers who understood the full risk profile of the industry as a whole".

He said that, while specialism in itself had advantages because it developed real experts in particular areas, a full understanding of the interrelationships between risks and risk management systems and processes was essential, particularly at the strategic level.

4.30 Mr Bird was concerned that present knowledge within the industry of the interfaces between what were now disparate companies could be lost. While the industry needed to adopt a unified approach to recruiting and training, cross-functional training was "hugely difficult" to achieve in the privatised industry. Companies now had different cultures, which affected their view and perception of training; they applied different criteria. Training was fragmented and there was a lack of common arrangements within the industry.

4.31 Even apart from any problems of privatisation, a concern of a number of parties was the loss of both industry-wide experience and specialist staff since privatisation. The Joint Rail Unions' view was that privatisation had led to a massive shedding of jobs, and, as a result of the loss of specialist staff, middle management no longer understood the day to day work of their staff. More training of first-line supervisors and managers was needed in areas such as safety leadership. Many new recruits to the industry, working in activities with safety implications, were insufficiently experienced or not properly trained. In particular, some of the driver training programmes were too short and unrealistic, particularly for those with little or no experience of the railways.

4.32   I should point out that the evidence in Part 1 of the Inquiry relating to Thames Trains' training of new drivers was to a similar effect. As Chapter 9 of the Part 1 Report demonstrates, there had clearly been a hiatus in the management of driver training since the days of British Rail, and a loss of "corporate memory". This had caused confusion over, and inconsistency within, procedures, at the very time when a large number of recruits new to the industry needed to be trained.

4.33   There was a very competitive situation in the recruitment of high quality skilled staff. Mr McClean said there was a risk that as stronger players developed in the industry, they would attract the better players. That could leave a number of industry parties short of direct skills, although the SRA's strategy of creating a smaller number of stronger players in the industry would deal with the problem in some respects. Mr J Knapp, General Secretary of the RMT, talked of

> "...situations now where companies poach workers from another...a company will offer £1,000 more or whatever to a skilled worker to go with them".

It was even suggested that in a situation where demand for skills exceeded supply, some companies might not think it worth training staff, if those staff were simply going to leave for jobs elsewhere.

4.34   These various difficulties with skills and training led Mr V P Coleman, Chief Inspector of Railways for the HMRI, to say that the industry needed to act now to increase the numbers of trained staff; it had got to decide itself how it was going to "bring on new people into the industry" and increase the cadres of properly trained and competent individuals. In Mr Baker's view the problems meant that training had to be improved on a cross-industry basis. Training schemes which were now company specific should perhaps be more broadly based, and there should be interchange of personnel. Training should be wider, and cross-industry training bodies should be further developed. Other witnesses identified the need for collaborative action, and mentioned the role of bodies such as ATOC, the Rail Industry Training Council (RITC) and the Institution of Railway Operators (IRO) in this context. ATOC, for example, had issued approved codes of practice relating to driver licensing, which covered items such as basic training and the transfer of safety performance information.

4.35   I see no need for me to make detailed recommendations about how the individual problems which were identified should be dealt with. However, the need for a skilled and properly trained workforce, at all levels of the industry, is clear. The various parts of the industry must take decisive action together to ensure that this need is met, and in particular that the difficulties caused by the fragmentation of the industry are overcome. I recommend that Railtrack and ATOC should work jointly with the RITC to set up a task force, with clear objectives and goals, which can review the issues and lead action to tackle them.

Large scale projects and the case for system authorities

4.36   There was considerable discussion during Part 2 of the difficulties that had been caused by the fragmentation of the industry in cases of the development and implementation of large scale projects involving a number of different interests. This is a subject which clearly embraces matters beyond safety alone. However it seems to me that safety aspects are so frequently critical to such projects that it merits close attention in this report.

4.37   The issue was highlighted in Professor Uff's report on the Southall crash, which found that the development of ATP had exposed difficulties in the management of cross-company projects, such as the lack of:

- a contractual framework governing the rights and obligations between the various players;
- an authority capable of instructing project managers etc; and
- a structure for the sharing and recouping of costs.

Professor Uff recommended that one or more system authorities should oversee the development of new projects and the continuation of work on existing projects, as well as one specifically to manage and fund the development of ATP.

4.38   Sir David Davies, in his report on ATP, supported the case for a system authority with the financial and operational powers to take a project forward. At the same time he acknowledged the practical difficulties for such a system authority that stem from the current regime of franchised operations, such as the risk of commercial disadvantages to individual companies concerned. The Joint Inquiry into Train Protection Systems also heard evidence on this subject. The report on that inquiry described the ad hoc industry liaison group which had been set up to take forward the train protection and warning system (TPWS), noting that it had no contractual or statutory status and had not been vested by its members with any power or authority. It concluded that a system authority without power to enforce its decisions would be of little value, particularly when faced with companies having different commercial interests.

4.39   In this Inquiry, the two favoured solutions were a single system authority responsible for all cross-boundary systems, and a series of system authorities each responsible for a single system. The latter had the greater support. Such systems would or could embrace commercial, as well as safety and technical, considerations. However, while the industry was already developing proposals for consultation under the leadership of the Rail Regulator, it was clear that a number of issues on which there was no clear consensus required further consideration. These included the role and remit of a system authority; whether there should be just one, or one per project; and its situation, constitution, and powers.

4.40   One view envisaged a single body rather than a multiplicity of authorities. Another saw system authorities as permanent bodies with an advisory role in regard to best practice, consistency, rules and Group Standards. However the predominant view favoured a series of time-limited bespoke bodies, established whenever a large scale project required the involvement of different parts of the industry. Their role would be proactive; it would embrace both the setting and the implementation of strategy over the life of a particular project, system or piece of equipment, from the initial stage of research and design, through development, to implementation and any subsequent modification. They would provide an authoritative framework within which decisions relevant to the system could be made, dialogue between regulators and operators could take place, and guidance on implementation could be formulated. A key objective would be to seek the best solution for the industry as a whole, rather than for any individual party to the project.

4.41 As suggested by Railtrack, the remit of a system authority would include considering proposed improvements; determining which industry parties were potentially affected and requiring their participation; determining funding arrangements; and initiating research, development and trials. Other tasks mentioned included the development of contract conditions, general project management and (although this might depend on the constitution and reporting line of the authority) the development and sharing of best practice. But it was clear that whatever its terms, the remit would need to be carefully determined and clearly set out, and gain the commitment of all interested participants.

4.42 This project-specific model of a system authority suggested that membership of the authority should be drawn from all parties needed to deliver that project, brought together in an alliance or partnership. The Joint Rail Unions suggested the importance of union representation.

4.43 Views on where, within industry structures, system authorities should be established were more diverse. The models advanced included system authorities established or overseen by Railway Safety, by a new technical standards body, by the SRA or by a new rail safety authority; overseen by a reconstituted SAB; formed by adapting the RISSC and the standards subject committees; or formed by agreement between (with reporting to) the parties concerned.

4.44 Other parties saw disadvantages with some of these models. For example, it was feared that, if system authorities were part of the regulatory regime, a lack of industry ownership would be created and legal powers would be required to enforce outcomes; while if leadership lay with the standards setting authority, that might compromise a system authority's independence. It was also suggested that the solution might depend on the project concerned. For example, an independent director might report to a national rail safety authority if the project concerned was safety driven, or to the SRA if the project was primarily operational or concerned with network strategy. The need for independence was stressed, although it was also argued that an entirely stand-alone body was not desirable. Transparent independence, moreover, could only be achieved if the majority of those forming the authority were appointed from outside the industry's parties, which might be difficult given the limited expertise available. The disadvantage of a solution which added unnecessarily to the proliferation of industry bodies was mentioned. Finally, depending on the model chosen, a means of appeal might be required.

4.45 The question of where and how system authorities were established was linked to that of the powers with which they would be provided. Again, no single solution was favoured. It was argued, on the one hand, that system authorities would require powers to bind all the players concerned and ensure their proper participation, including financial participation. The source of the system authorities' powers was thus seen as a key issue, since these players might well include bodies other than duty holders. It was equally argued, however, that to give system authorities separate powers or obligations might unacceptably dilute the responsibilities of duty holders, or create ambiguity in their roles.

4.46 One model which was being actively considered by the industry for the establishment of system authorities was to use a Railway Group Standard, with management of the system authority lying with the standards body. This method was seen as providing both power and impartiality. Alternative means of establishment, under current arrangements, were proposed, including licence modifications and track access conditions, but it was suggested that these might suffer from the disadvantage of not providing adequate powers over some of the bodies, such as ROSCOs, which might need to be involved. It was also suggested that, depending on the model chosen, statutory backing might be required, whether to provide for the authority's constitution or to ensure that it had sufficient powers – for example, in situations where the optimal solution overall did not favour each participant equally.

4.47 Arrangements for funding system authorities were also seen as depending on the form of constitution chosen. For example, system authorities established under a rail safety authority could be funded from the levy for that authority. Stand-alone system authorities could be funded directly by those participating in the project concerned, as part of the project costs.

4.48 I welcome the industry's clear acceptance of the case for system authorities, and the consideration that has been giving to their development under the leadership of the Rail Regulator. As is clear from what I have said above, no clear consensus emerged in the Inquiry as to how they should best be constituted. I do not think it necessary, however, for me to make detailed recommendations in this respect, not least given that the Inquiry was told that the S&SD (now Railway Safety) were developing a Group Standard to provide for their establishment, and had issued a discussion paper for consultation in November 2000. I believe it is important that this work continues. However it is also important that as these proposals develop, proper account is taken of the views expressed to the Inquiry. The solutions chosen must be those best able to deliver the main objectives of system authorities. Although overall leadership will be important, system authorities will need to be properly empowered bodies, not always appealing to a higher authority. They must themselves provide clear leadership for the project concerned, and be able to ensure the commitment of all parties to their work and their decisions. Further thought will be needed about how these decisions can be properly enforced, if this proves necessary. They must also have the means of ensuring that they have the finance which they need, through proper and equitable contributions from participating bodies. Finally, in order that responsibilities are not confused or duplicated in the longer term, it is desirable that system authorities do not remain in existence for longer than is justified by their particular task. In due course, responsibility for the project under development will need to pass to one or more permanent bodies.

Research and development

4.49   The pursuit of research and development within a fragmented industry was considered to present similar problems. Prior to privatisation railway research and development within the UK were predominantly carried out by British Rail's Research Department at Derby, while funding came from the British Railways Board. Following privatisation this research group was broken up, and its constituent parts became commercially-orientated businesses. Mr Muttram suggested in his written evidence to the Inquiry that this fragmentation could lead to a decline in research projects in different parts of the industry.

4.50   In part to guard against this possibility, a working group was established in 1996 under Sir David Davies to examine research in the rail industry and to propose how it might be taken forward. Sir David's 1996 report recommended the establishment of a small group able to take a strategic view of the future of the railways and sponsor research on topics which spanned different parts of the rail industry. It recommended the formation of a railway research association which could sponsor research as required. In the event no such collaborative body has been established.

4.51   In his report on ATP Sir David Davies further commented on the lack of collaborative research on key issues of strategic importance to the industry, such as traction, current collection, braking, signalling, train control and other safety issues. He pointed out that these predominantly fell outside the sole responsibility of any single body within the industry. In his oral evidence to the Inquiry Sir David stressed that research needed to be done. In his report into the Southall rail accident, Professor Uff also emphasised the need to put in place means to resolve inter-company issues relating to research and development at all levels. He recommended stringent control of research and development processes with regard to their programming, cost management and funding arrangements, and he concluded that funding should be on a cross-industry basis.

4.52   Mr Muttram also referred to the lack of co-ordination, direction and collaborative funding of research projects. In his evidence to the Inquiry Sir David Davies said that Railway Safety should be responsible for safety research. This was supported by Mr R J Morris, Executive Director of London South East and formerly the Technical Director of Safety and Operations for the SRA. Mr Morris said that Railway Safety would hold the budget and be the instigators of safety research. There was universal acceptance by the parties to the Inquiry that funding for research should be on a cross-industry basis, probably raised via a levy on the companies. I should add that I consider that there is merit in the submission by ATOC that there is a need for a database which would enable members of the industry to find out the location and ownership of relevant research material.

4.53   This general situation on research and development was contrasted with the situation in Japan. In the seminar held by the Inquiry on the Japanese Model of Rail Safety Professor R I Smith, Head of the Department of Mechanical Engineering at Imperial College and Chairman of the Advanced Railway Research Centre, spoke to a report of a rail industry group visit to Japan in early 2000. The group were impressed by the Railway Technical Research Institute at Tokyo, which is the major railway research centre for the world and has been in existence for almost a century. It is staffed by some 600 people, with 80% of its funds coming from the Japanese railway companies on a levy basis based on their traffic revenues.

4.54   It is clear that rail industry research in the UK falls below international standards. This situation should be redressed urgently. I agree with Sir David that research and development should be led by Railway Safety, with the SRA and the Rail Regulator providing support as necessary. Further funding should be based on a levy on the participating bodies in proportion to their railway-based income. I will, however, require to consider in Chapters 8, 9 and 10 whether the role played by Railway Safety should be taken over by another body.

The franchising process

4.55   The franchising process was considered by witnesses to have disbenefits for safety. These arose not only from the number of franchises, but also from the short length of the original franchises and from a lack of emphasis on safety within the consideration of franchises.

4.56   As a result of the privatisation of the rail industry TOCs were franchised to operate on the rail network for seven years. Mr Corbett told the Inquiry that both he and Mr C Green, Chief Executive of Virgin Trains, felt there were too many franchises in the industry. This point was re-stated by Sir Alastair Morton, who referred to the SRA's policy of having fewer and stronger franchise owners. He said:

> "That may translate into fewer, more rational and therefore stronger or easier to operate franchises. But it is the financial and managerial muscle available to any given part of the railway that we are aiming at. Fewer and stronger in that sense to stand up to the ebbs and flows of financial and operating good and bad fortune".

4.57   Sir Alastair went on to discuss the length of time for which a franchise should run. When asked whether he considered that seven years gave little incentive for a franchisee to invest heavily in safety unless he could be confident that he was going to get that investment back in one form or another, he agreed that seven years was a short time. He said it was now recognised that franchises should be granted for a much greater duration, of from 15 to 20 years. The intention was to use the longer franchise to obtain a commitment from the operator to drive up safety standards and enforce safety compliance. This would allow operators to realise such safety commitments and profit from them.

4.58   Mr Morris told the Inquiry that the process of lengthening the period for which a franchise should run had commenced and that the most recent franchises let were for a period of 15 to 20 years.

4.59   Irrespective of the length of the franchises, Railtrack were clearly concerned that they should have an effective safety input into the franchising process. They submitted that since they had responsibility to ensure that risk was not imported on to the infrastructure, they must have the right to control avoidable importation of risk by appropriate standards and controls. Thus, at the very least, they should have the right to be consulted on the ability of any particular potential franchisee to manage safety. An operator's safety case alone was insufficient, in their view, to ensure that the future franchisee had the right safety ethos to manage safety on the railways properly.

4.60   The SRA clearly accepted that in the re-franchising process they should consult Railtrack and the HSE on matters that could have a bearing on safety. In the current round of re-franchising there was a "very clear process" whereby they involved both bodies when considering the safety part of bids. Mr Morris confirmed that the SRA would take the S&SD's advice into account both when short-listing candidates and before deciding on the preferred operator status. He also said that there would be a memorandum of understanding between the SRA and Railtrack in this regard. Mr Muttram welcomed this as "a major step forward". It is important that the HSE are consulted on all aspects of a bid that may have safety implications and not just those which deal explicitly with safety. I return to this subject in paras 9.99-9.102.

The use of contractors

4.61   The Inquiry heard a considerable body of evidence regarding the employment of contractors in the rail system, much of it in criticism. The use of contractors is most notable in the case of Railtrack and the evidence concentrated on this. In a paper to the Safex Committee of Railtrack dated 9 October 2000, Railtrack's Assurance and Safety Director Mr J Abbott, describing the historical background in the work of contractors, said:

> "Previously maintenance and renewals were undertaken by internal labour and the use of contractors was restricted to selected project work. Now Railtrack contracts for all its maintenance, construction, renewal and design".

Railtrack suggested in their statement of case there was no ground for believing that their reliance on contractors in any way prejudiced safety. The evidence to the Inquiry, however, called this assertion into question.

4.62   A problem with regard to the competency of staff engaged by some contractors and subcontractors was identified by a number of witnesses. Counsel for the Joint Rail Unions drew attention to another passage in the same paper. It described:

> "...an industry within an industry supplying both safety-critical and ordinary labour of which there are about 130 suppliers and a number of new applicants.
>
> It become *(sic)* apparent that abuses of the systems in place were rife with allegations of inadequate training, exceedance of working hours and lack of any safety systems within the supplying companies".

4.63   Individual problems mentioned included a falling-off of standards of training; the unfamiliarity of contractors and their staff with the railways; training sessions which were too short; a lack of continual assessment by contractors of their staff; a failure by Railtrack to ensure that contractors' staff were properly trained; and concern as to the rigour and standards of some external training companies.

4.64   Evidence was given to the Inquiry about an evaluation carried out by the HSE in 1996 of Railtrack's safety management systems. Their report "Maintaining a Safe Railway Infrastructure" concluded that there was nothing to suggest that the Safety Case Regulations, or Railtrack's safety case, were deficient. The HSE noted, however, that in practice the infrastructure contractors did not always comply with Group Standards. It also found instances of Railtrack staff agreeing with contractors to non-compliant procedures.

4.65   Mr Coleman observed that DuPont Safety Resources had made criticisms in their analysis of the safety management process in Railtrack in 1999, which were "similar in kind to those made of Railtrack in the 1996 report".

4.66   The DuPont report, which followed a very wide and thorough investigation of Railtrack's safety processes, stated with regard to the safety aspect of contracting:

> "The expectations for the safety of the contract workforce should be identical to those of direct employees. This also relates to the safety performance standards that are set and followed. Contractors also provide a critical link in the system for ensuring high standards of passenger and public safety.
>
> The contractor safety management should include elements of:
>
> - pre-qualification
> - planning the work and agreeing the expectations
> - induction and training
> - control of work practices to the appropriate degree
> - review of performance".

4.67   The DuPont report, which was accepted by Railtrack, commented with regard to contracting:

> "The multiple operational interfaces between Railtrack, the TOCs and contractors are vitally important to performance and to the safety of the public and employees. We have found these interfaces to be weak and enforcement of contracts to be poor or non-existent. In many cases, compliance with safety rules is poor, putting contractor employees at risk".

4.68   DuPont went on to point out that Railtrack audits found that only about 60% of contractor work was compliant with safety case issues. They also pointed out that, due to the lack of authority of most of the auditors, the audits of contractors' performance were limited in their effect, and that the auditors reported that the training of the employees of infrastructure contractors was poorer than in the past. The auditors, engineers, and contract managers reported many failures of contractors to comply with standards and operating procedures. When these points were put to Railtrack they suggested that the DuPont report described a system and situation which existed under the older form of contract (RT1A). Mr Corbett accepted that this form of contract created difficulties. He described RT1A as an adversarial contract. He went on to maintain that under the new contract (IMC 2000) which was currently being introduced and which was "a partnering contract open book based on target costs" these deficiencies had been addressed. Since this was very recent the Inquiry heard no evidence as to how well IMC 2000 was addressing the problems, but Mr Corbett did go on to agree that, although a new contract system was necessary, "the answer to deep-seated difficulties will not lie in better contracts". Inevitably the contracts created a different set of interfaces. He said:

> "There is the risk, and I say it is only a risk, that different safety cultures can emerge in different parts of the system".

4.69   I will address the evidence presented to the Inquiry regarding these problems in the use of contractors in four areas:

- the process for the award of contracts;
- the control of contractors' and sub-contractors' work;
- the control of contractors' and more particularly sub-contractors' staff; and
- the size of the contractor and sub-contractor population.

The process for the award of contracts

4.70   Mr C J Wheeler, Project Manager of the National Track Safety Strategy Group and Chairman of the Association of On-Track Labour Suppliers, and the Joint Rail Unions were critical of the process employed to award contracts. Mr Wheeler was critical of the time available for preparatory work following the award of a contract. He told the Inquiry that it was not unusual for a contract to be let on a Tuesday for work to begin on a Saturday, thus providing little or no period for preparation. This was a particular problem when sub-contractors had to be hired in the interim. The Track Safety Strategy Group had notified Railtrack of this problem. They had "made the right noises", but had not yet taken "the right action".

4.71   Mr Knapp spoke of his concern regarding the short length of time for which contracts ran. It was unacceptable for the main contractors to have to re-tender for their work every five years. This required senior managers to spend an inordinate amount of time in defending their contracts. Mr Wheeler also stated that the longer a contract ran, the better was the safety performance. The Joint Rail Unions also suggested that there were problems in the arrangements made for allocating responsibility for designing and planning the work given to contractors. Mr Knapp mentioned "men being hired on a Friday night" for immediate work as sub-contractors' labour without receiving adequate training. Mr Coleman spoke of the need for "improvement in all of these areas".

4.72   All of these points lead me to conclude that the current process for the award of contracts is not being operated in an appropriate manner. It is essential that adequate steps are put in place to ensure that contractors and sub-contractors are selected by a process which gives due regard to their state of training, and they are given appropriate time further to develop their training and planning as necessary before embarking on any work.

The control of contractors' and sub-contractors' work

4.73   Railtrack seek to control the quality of a contractor's work by requiring the submission, as a condition of contract, of a contractor's assurance case. It was obvious to the Inquiry that this control was ineffective. I have previously referred to the DuPont report regarding Railtrack's own audits of their contractors where they found about 60% compliance with safety case issues. It is obviously unacceptable that 40% of the work being performed by contractors is non-compliant. When I consider that in many cases Railtrack rely on contractors to check other contractors' work, then the cascading effect of this non-compliance is of particular concern. The process of contractors checking other contractors was described by Mr D W Wilks, Infrastructure Contracts Manager for Railtrack Southern Zone. He said that the weekly inspections required by controllers, the eight-weekly inspection required by a section manager, the annual inspection of wheel timbers by a bridge examiner and the two-yearly examination required by a permanent way maintenance engineer were all conducted by contractors. With regard to the future, Mr Wilks explained that contractors would still perform these functions after the introduction of IMC 2000.

4.74   I also heard of concerns regarding the quality of the training given to contractors. The Collins Passengers' Group, in their statement of case, reminded the Inquiry of a report by the Environment Transport and Regional Affairs Committee of the House of Commons in November 1998 which concluded that Railtrack needed to tighten their procedures for training their contractors. As I have previously noted, the DuPont report recorded that the quality of the training of contract employees had deteriorated.

4.75   I conclude, therefore, that the controls in place for the management of contractors' and sub-contractors' work are inadequate. It is essential that steps are taken to ensure that the quality of the work carried out by contractors and sub-contractors entirely meets the required standards, and that any deficiencies are addressed in a timely manner.

The control of contractors' and sub-contractors' staff

4.76   It is obvious from the foregoing analysis that there is a need to control contractors to ensure that they employ only adequately trained staff. Steps need to be taken to ensure that the staff adhere to the working hours as prescribed by Group Standard GH/RT 4004. The National Competency Control Agency record qualified individuals and issue "Sentinel" cards. The Sentinel system provides a register of some 100,000 employees from around 2,000 companies, which can be accessed by phone using a magnetic card. The problem lies, however, in the industry's inability to enforce the standards and to prevent abuse of the control system. Mr Wheeler gave as an example to the Inquiry how one Sentinel card holder could be sponsored by up to seven separate organisations, with no one company knowing how many hours the employee had worked for the others.

4.77   Problems also exist when a railway company employee, or even someone normally employed outside the rail industry, works in parallel for other infrastructure maintenance contractors or sub-contractors. Mr C Carr, Technical Director of Amey Rail, described how adherence to the requirements on the rescheduling of working hours, recommended by Sir Anthony Hidden in his report on the Clapham crash, was difficult if not impossible to police if employees also worked for other organisations.

4.78   In the paper presented to Safex by Railtrack's Assurance and Safety Director Mr
        Abbott, to which I referred in paras 4.61-4.62, he acknowledged this problem stating:

> "Current intelligence is, despite the 'beefing up' of the management regime
> through Sentinel, that there is systematic abuse with working hours being
> exceeded (regularly rather than by exception), rest hours inadequate, excessive
> shifts being worked and a fundamental lack of route knowledge. Although the
> evidence can be difficult to establish, it would appear to be widespread and
> growing".

4.79   Considering these problems, I conclude that there is a need for an immediate and
        sustained improvement by the industry in the manner in which they control
        contractors' and sub-contractors' staff. I support the recommendations suggested by
        the Joint Rail Unions that contractors and sub-contractors should all be registered
        through the National Competency Control Agency, and that Railtrack should monitor
        this process to ensure that the contractors they employ are adequately trained and not
        working excessive hours.

4.80   I further recommend that the Sentinel system be reinforced in its application across the
        industry with specific attention being given to the need to record the hours that any
        individual works on the railways and to ensure that the Sentinel card is clearly "tied"
        to an individual, perhaps by incorporating a photograph, as part of the identification
        process.

The size of the contractor and sub-contractor population

4.81   Mr Wheeler pointed out that the fact that there were 2,000 contracting companies
        engaged in the rail industry, between them employing some 100,000 staff, created a
        significant number of interfaces. He stressed the absence of any mechanism to bring
        the contracting companies together. Mr Corbett told the Inquiry that he believed that
        the number of contractors had to be reduced because the contract areas covered were
        too small to encourage the necessary investment. He went on to say that it might be
        necessary to "think the unthinkable" and take some of the contracted tasks back under
        the direct management of Railtrack.

4.82   I support the argument for the reduction of the number of contractors. It is clear that
        the industry has been unable properly to control and manage the work of its
        contractors, and by having a smaller population of contractors, each with a larger area
        of responsibility, control could and should be improved.

The management of contracting

4.83   Railtrack should take a direct and active role in the close day to day management of
        safety-critical work. I welcome the steps that have already been taken in that
        direction.

4.84   The new contract system, IMC 2000, will encourage a necessary move towards more
        collaborative arrangements and long-term partnering.

4.85	In Chapter 7 I will discuss the proposal that contractors should be required to hold a statutory safety case.

4.86	One way of encouraging long-term partnering with contractors was described to the Inquiry by Mr J T Atkinson, Manager of Rail Safety for the Land Transport Safety Authority of New Zealand. He explained how he saw contractors as part of the labour force of the licensed infrastructure. The Inquiry heard a similar view from Dr I A P Scott, Director of Safety, Health and Environment for Eurotunnel, when he described the operation of the Eurotunnel Contractors' Committee. This had as one of its purposes to

> "…encourage contractors to adopt the same attitude to safety, the same type of safety rules and the same adherence as we ourselves were doing".

In the same vein, Mr D Tunnicliffe, formerly Chief Executive of London Transport and Chairman of LUL, spoke about the need

> "…to work with your contractors to manage the special risk of the railway environment. It is much more important to make sure they have systems in place to manage the joint risks…".

4.87	I support the concept that contractors and sub-contractors should work to exactly the same safety standards as those employed directly by the industry. It is the responsibility of the employer, in this case Railtrack, to ensure that this is done. The key to this must be ensuring the competence of those engaged on the work. However the same applies to TOCs in so far as work is done for them by contractors and sub-contractors. Mr Corbett raised the possibility of a training school for contractor staff. I endorse this proposal and I suggest that Railtrack should seriously examine the possibility of such a training school as a matter of high priority. It may well be that they might find the frameworks already established in the National Vocational Training Schemes to be worthy of consideration in developing this proposal.

4.88	Finally, I note that the HSE have wide powers to approve assessments under Regulation 3 of the Railways (Safety Critical Work) Regulations 1994. If concerns persist regarding the competency of sub-contracting staff, they should not hesitate to use these powers as required.

The role of the trade unions

4.89	Privatisation undoubtedly brought about a significant change in the role and influence of the trade unions. The Inquiry heard views on this subject in the seminar on Employee Perspectives on Rail Safety and the seminar on Developing an Effective Safety Culture. Witnesses representing the Joint Rail Unions also submitted significant evidence to the Inquiry.

4.90	Under the Safety Representative and Safety Committee Regulations 1977 a recognised trade union may appoint safety representatives from amongst the employees, and those representatives have certain specified functions. The regulations impose a duty on the employer to consult the safety representatives in good time on certain specified matters. The Health and Safety (Consultation with Employees) Regulations 1996 provide for consultation where there are employees who are not represented under the 1977 Regulations. I note that in November 2000 the responsible Minister agreed that the HSC should take forward certain proposals which included:

(i) giving the employees in non-unionised workplaces the right to decide on the method of consultation; and

(ii) the running of a pilot scheme for "workers' safety advisors".

The Joint Rail Unions suggested that safety representatives on a site should have the right to represent the staff of contractors at that site. However, as this may raise issues which go well beyond the case of those employed on the railways, I do not consider that I should make any observations about it.

4.91   The Joint Rail Unions complained that little had been done to enforce these regulations. It was suggested at the seminars that there was an anecdotal basis for the view that, perhaps due to operational pressures, safety issues raised by the trade unions were given lower priority than in the nationalised industry.

4.92   The seminar on Employees Perspectives on Rail Safety was told by some employees of their perception that post-privatisation, trade unions were less involved. One safety representative, who was also a trade unionist, said that he felt "a sense of alienation". The seminar was also told that in some instances safety briefings were being delegated to the trade union safety representative. It was agreed at the seminar that that is not an appropriate role for a safety representative. The role of line management in communication is a matter to which I will return in Chapter 5.

4.93   The seminar was also told of the difficulty of recruiting volunteers to be safety representatives. It was claimed that trade union representatives could be seen as a "nuisance factor", and that this was an inherent problem of the railway culture. This appears to betray a lack of trust. It was stressed at the seminar that safety representatives in general and trade unions in particular have a major part to play in the safety process, particularly as they have a detailed knowledge of the work being performed and of the safety aspects pertaining to that work. I will return to this question of trust in Chapter 5.

4.94   Many of the points raised in the seminar on Developing an Effective Safety Culture echoed those from the seminar on Employee Perspectives on Rail Safety. In addition, it was claimed at this seminar that some people considered trade unions to be an undervalued resource. They were an important conduit for employee views, and could provide a "bottom up" audit. They could facilitate employee feedback. They provided a source of independent and alternative expertise, and they gave it anonymity which was a source of confidence.

4.95   There were also some positive comments; for example, that the trade unions were now being involved in changes to Group Standards, and were invited to join in working parties on human factors. I consider that, while their membership of suitable bodies should be kept under review, there is no need for me to make a recommendation as to the existing bodies on which they should be represented.

4.96   Mr Knapp, Mr R Rosser, General Secretary of the TSSA and Mr M Rix, General Secretary of ASLEF, expressed their collective concerns in a number of areas. They wanted to see more trade union involvement in the early stages of the preparation of safety cases, where they believed they could contribute specific operational information (see para 7.25). At present, they said, union involvement came at the end of the process when most matters had already been settled. They felt that following privatisation it was difficult to raise cross-company issues, since safety representatives and safety committees now related to a single employer only, and the central groups which existed under British Rail had been disbanded.

4.97   They had concerns about the level of training in the industry, particularly in the area of training new employees, and volunteered to assist in that process. As I have already noted, they expressed concern about the level of training of sub-contractors.

4.98   They asserted that there was still a culture of performance over safety and claimed as examples that drivers were being encouraged to disregard defensive driving techniques in locations of low adhesion in order to keep up with the timetable, and that safety briefings for employees were being cancelled because of production requirements. While these matters give rise for some concern, I believe that they are problems which can and should be dealt with by the normal day to day relations between management and the unions, and that there is nothing specific in the set-up of the privatised industry that is causing or exacerbating these problems.

4.99   I would, however, stress that it is the responsibility of management to ensure that the elected representatives of the employees, whether they are union officials or not, have a significant role in the management of safety. Management should recognise and act on this.

4.100   Not all railway employees are represented by trade unions, although approximately 70% are. The DuPont report described how it is possible to create an open discussion platform with employees without having to rely on third party representation. This enables direct communication between all levels of the organisation. I will return to this subject in more detail when I discuss communications in Chapter 5.

4.101   While there has undoubtedly been a change in the way in which trade unions relate to the fragmented rail industry, this has not of itself had a detrimental effect on safety. Consultation and the resolution of disputes can and do take place at a local level appropriate to the needs of the situation and avoiding problems reaching the national level.

Chapter 5
The management and culture of safety

Introduction

5.1 The central importance to the Inquiry of the effective management of safety in the rail industry was underlined by a number of witnesses, including Mr G M N Corbett, then Chief Executive of Railtrack; Mr J Kooger, Senior Consultant for DuPont Safety Resources; Sir Alastair Morton, Chairman of the SRA; Dr I A P Scott, Director of Safety, Health and Environment for Eurotunnel; Mr J W Smith, Head of Regulation in Railtrack; and Mr D Tunnicliffe, formerly Chief Executive of London Transport and Chairman of LUL. In the words of Mr Tunnicliffe:

> "If the Inquiry solely works on the issues of structure it is perfectly possible for it to create a new structure, but if it has not somehow or other engaged the industry in safety leadership the net impact on safety may well be modest".

The Inquiry heard various opinions as to what constituted the essential elements of successful safety management. It also commissioned two reports from Entec UK, on Safety Management and Safety Culture, and on Accident Causes and Responses. It is neither possible nor appropriate for me to attempt to draw from them a definitive view. However, in this chapter I consider the most important issues to emerge with regard to the industry's management of safety, and the safety culture within which that management is carried out. The subjects covered are:

- the role of safety management and safety culture (paras 5.2-5.10);
- safety leadership across the industry (paras 5.11-5.17);
- safety leadership within individual companies (paras 5.18-5.24);
- communications (paras 5.25-5.37);
- staff motivation (paras 5.38-5.40);
- toleration of unsafe acts and the "blame culture" (paras 5.41-5.50);
- continuous learning (paras 5.51-5.56);
- training and competency (paras 5.57-5.63);
- "interdependency" (paras 5.64-5.67); and
- conclusions (para 5.68).

The role of safety management and safety culture

5.2 Mr V P Coleman, Chief Inspector of Railways for the HMRI, subscribed to the HSE's view that accidents, ill health and adverse incidents are seldom random events. The immediate cause may be human or technical failure, but these in turn usually stem from organisational failures which are the responsibility of management. Mr Kooger emphasised this point when he told the Inquiry that people behaviour was one of the most important factors in safety management. In his opinion 90% of all accidents in any company were likely to be due to deviations in behaviour rather than to functional failures, such as failures in systems or equipment.

5.3 All the relevant accidents (see Appendix 4) highlight deficiencies in the management of safety in some way or another. This was the conclusion drawn by the HSE in their statement of case, and supported by Entec's report on Accident Causes and Responses. The relevant accidents provide examples of problems associated with people and management behaviour, whether it be, for example, acts of omission or commission; poor decision-making; poor communications arising from a lack of clarity as to responsibilities and accountabilities; conflicts between safety and performance; failure to identify risks and develop controls; poor follow-up of recommendations; and lack of training and competency.

5.4 The clear conclusion is that a high proportion of accidents, incidents and near misses on the railways follow unsafe acts by people, whether front line workers or managers. I do not seek here to imply, by any means, that all such people are performing their duties below the standard which should reasonably be expected of them. They may very well be dedicated employees, working to the best of their abilities. Rather, their unsafe acts should be seen as the result of underlying deficiencies in the management of safety, and tackled accordingly. It is in this light that the management of safety was considered in the Inquiry.

5.5 A fundamental factor in the management of safety is the safety culture which prevails, whether by accident or design, both across the rail industry as a whole and within individual companies. In the submission of the HSE:

> "…the need for a positive safety culture is the most fundamental brought before the Inquiry".

Mr Corbett referred to leadership and culture as key management tools in the improvement of safety, while Mr Kooger emphasised that the proper management of safety required a proper safety culture.

5.6 The Inquiry heard various definitions of the word "culture". Counsel to the Inquiry submitted that a good safety culture was:

> "…the product of individual and group values, of attitudes and patterns of behaviour that lead to a commitment to an organisation's health and safety management. Organisations with a positive safety culture are characterised by communication founded on mutual trust, by shared perception of the importance of safety and by confidence in the efficiency of preventative measures".

5.7 Many of those who spoke on the matter said that safety culture could not be separated from the wider culture of a business, and that a reliable view of it should not focus on safety alone but rather on the delivery of the business as a whole. The more management focussed on the delivery of safety as an integral part of their business, the more likely they were to succeed. Thus Mr McClean, Production Director of GNER, described his company as making safety a visible element of the management process, while in the words of Mr C J Wheeler, Project Manager of the National Track Safety Strategy Group and Chairman of the Association of On-Track Labour Suppliers:

"…at board level safety is generally recognised as a top priority which is good for business".

5.8   The consensus among the participants in the seminar on Developing an Effective Safety Culture supported the view that culture is a reflection of the overall attitude of every component of management within a company. Mr Mogford, Director of Health and Safety in BP, put it thus:

"The more you drive to deliver safety as an integral body of business the more likely you are to be successful. I do think you need to talk about it in a business context and not just in terms of safety".

5.9   There was, therefore, a general view that a desirable safety culture would be business-driven but with safety as the highest priority. All decisions taken would be considered for their safety aspect; management and employees would strive to deliver a high level of safety; compliance with the safety rules and standards would be the norm; and employees would accept their role within the framework of the rules and procedures. They would not only conform to these rules and procedures themselves but would also use their best endeavours to ensure that their co-workers were working safely. It is clear, therefore, that there is a link between good safety and good business. The rigour and operating discipline of the processes which lead to good safety performance are exactly the same processes which deliver good operational performance, which in turn leads to good business. There is no "trade-off" between safety and good business. As was said by the Rail Regulator: "…they are two sides of the same coin".

5.10   It was suggested that the Inquiry should consider the proposal that the effective management of safety within the rail industry would be facilitated by the establishment of a compulsory mutual insurance fund. However I decided that, while this issue may warrant further consideration by the industry, it was not appropriate to embark upon an investigation of it within the context of the Inquiry.

Safety leadership across the industry

5.11   The first priority in striving for a successful safety culture must be leadership. As Mr Tunnicliffe observed: "I simply overwhelmingly believe that leadership is what gets things done". There are two aspects to consider: leadership across the industry, and leadership within individual companies.

5.12   The report by DuPont Safety Resources which was produced in January 2000 for Railtrack, stated that leadership was at the core of any attempt to improve safety management on the railways. The first area of concern identified in the report was that although various bodies and individuals played a part, there was "no clear identification of safety leadership in the UK Rail Industry".

5.13   These views were reinforced by participants in seminars organised by the Inquiry. The seminar on Developing an Effective Safety Culture demonstrated a concern that the existing culture within the industry was characterised by, amongst other things, a lack of leadership. It was moreover bending under pressures such as those from the regulators, and suffered from problems including the fragmentation of the industry, the blame culture and confusion over responsibilities and accountabilities. The seminar on Public Perceptions of Rail Safety expressed concern at the absence of a coherent co-ordinating body for the railways, and suggested that the public did not understand the various responsibilities and accountabilities within the industry. These and other problems contributed to the public's feelings of unease about the railways and a decline in trust and faith. Some of those attending the seminar on Employee Perspectives on Rail Safety considered that the large cultural change across the industry which had followed from privatisation had meant that there was no longer any interlinking of culture between companies.

5.14   The consequences of failures in safety leadership were highlighted by a number of the relevant accidents. For example, the HSE report into the accident at Watford South on 8 August 1996 noted that

> "...most of the recommendations required Railtrack, as the infrastructure controller, to take on a pro-active and co-ordinating role so that they can be satisfied that risk is being properly co-ordinated on their infrastructure by the train operators and others".

In the case of the Bexley accident on 4 February 1997, the HSE concluded that Railtrack had failed to implement an improved strategy for monitoring contractors' performance. This can be seen as an example of a lack of leadership leading to a failure to tackle a recognised problem to ensure that an effective and safe system is implemented in practice and not just on paper. Perhaps the most striking example of a failure of leadership was shown by the history of the ATP pilot project. In the words of the Joint Rail Unions' statement of case:

> "...the history of the GWT ATP pilot throws into sharp relief the conflicts in a disaggregated industry...There was clearly no proper and clear project management and the pilot was allowed to limp on and should have been brought to fruition much earlier and the lessons learned".

Professor Uff in his report on the Southall crash came to the conclusion that the problems with the ATP pilot could have been resolved

> "...within a very much shorter time span had there been greater commitment and allocation of resources in the period before and following privatisation. The delay which occurred can be explained but not excused. GWT bear a major responsibility for the delays, but the actions of BR, Railtrack and HMRI all played their part...and the absence of any co-ordinating system or authority was pivotal".

5.15   A number of witnesses reinforced the view that fragmentation had made it more difficult for the industry to think and act on safety in a unified way. Mr Coleman, for example, told the Inquiry that the greater the fragmentation, the greater the challenge to the achievement of leadership and co-operation. He saw problems of defensiveness and resistance to change, and stressed that any improvements would have to come from within the industry itself. Mr Corbett, who recognised leadership and culture as key management tools, agreed that an overall vision was required for the industry, but spoke of a diffusion of accountability which in the past had made it unclear who was expected to give safety leadership.

5.16   At the same time the Inquiry found widespread recognition that the industry had accepted the problem of leadership and was beginning to take action to solve it. A need had been recognised, for example, for a body to take a lead role and for the individual leaders of the constituent parts of the industry to collaborate with mutual respect and in a transparent manner. Perhaps inevitably, it was less clear where the responsibility for leadership should lie. Mr Corbett felt that it was a task for himself as Chief Executive of Railtrack, and stated that that was why he chaired the National Safety Task Force. Mr R I Muttram, Director of the S&SD, agreed that the infrastructure controller would always have a vital part to play in this field. However he also considered that Railway Safety should be leading the industry in its safety improvement programme, and that this could be achieved through mechanisms such as the Railway Group Safety Plan. Mr McClean agreed that this plan was having some success in encouraging different parts of the industry to recognise common objectives and co-operate. Others, too, agreed that the Railway Group Safety Plan was improving cohesion, even though some had found that its implementation was poor and its aims were lowered. Further signs of improvement were found in the new sense of direction provided by the Government's 10 Year Plan for the industry and by the developing role of ATOC in representing the interests of train operators.

5.17   In my view, there is a need for a rail industry body which can take the leading role in the promotion of safety. I return to this subject in Chapter 10. This would provide the leadership and structure which are needed, and should go a long way to improving the clarity of operation. However, it will still require all the component companies of the industry to accept and support this leadership. This is a matter well beyond structures. It will call for a shift in behaviour. The success or otherwise of the behavioural change which is required will be fundamental in whether or not the railways will achieve the demanding safety goals now sought for the industry.

Safety leadership within individual companies

5.18   Clear and decisive safety leadership is equally required within individual companies. Mr Kooger, for example, spoke of strong, demonstrated management commitment as the "basic component" of a successful safety programme. In his view:

> "...in order for safety to be considered as another key parameter, that type of conviction needs to come from the top, needs to be broadcast from the top, needs to be continually refreshed from the top and the top needs to be seen, visibly seen as active in its representation of its value for that parameter...It means that the subject is repeatedly and continuously brought to the attention of everybody. It is repeatedly mentioned in every bulletin and communication. The chief is actively seen as being interested in the subject and being committed to it".

5.19   Many examples of such leadership in the industry were brought to the Inquiry's attention, but it was also evident that much more required to be done. Mr P J Waite, Technical Director of Entec UK found that the commitment from directors or chief executive officers was variable, with both good and bad examples to be found within the safety cases which he had examined.

5.20   Counsel for the Joint Rail Unions quoted the report by DuPont on the management of safety within Connex Rail, which stated:

> "The commitment to safety of senior executives is not visible at the working and operational level and the safety policy lacks credibility in the eyes of the employees".

5.21   Given the remit and nature of the Inquiry it is not appropriate for me to recommend actions to be taken by particular companies. However, I can make general recommendations, against which individual companies should assess their own performance. Above all, Chairmen and Chief Executives of companies must make continually clear to all their employees and passengers a lasting commitment to improve safety performance. The success of this process should be judged from the position of employees and the travelling public; that is, what they see and hear from the senior executives.

5.22   Dr Scott told the Inquiry of the programme of regular walkabout visits to which Eurotunnel's senior management team was committed. He said it was a powerful tool for convincing staff that the directors were interested in safety, for getting the message across, and for listening. For some in the rail industry the benefits of such an approach were clear. Mr Bird, Chairman of c2c Rail, spoke of

> "…selling the concept that safety matters. The whole management team is out there understanding as to whether the safety instructions are actually understood and being applied".

Mr Corbett, speaking of his experiences before and in the wake of the Hatfield crash, said:

> "Sitting where I sat, when you look at the data coming towards you, none of our safety KPIs (Key Performance Indicators) were going in the wrong direction other than trespass and vandalism…However, in recent months, getting out and about and you hear the signallers talk and you meet some of the drivers and you go out with the teams on the track, and then with what happened at Hatfield, I think one does have to question whether the noise level at the front line actually has generated a set of behaviours which we did not know about and did not intend and whether we do have that as an issue".

5.23   Many of those who attended the seminar on Employee Perspectives on Rail Safety considered that senior management had poor visibility and were rarely seen unless on VIP visits. There is no substitute for personal contacts. Mr Kooger spoke of spending one hour of each day touring the workplace and talking to the employees about safety. Companies in the rail industry should be expected to demonstrate that they have, and implement, a system to ensure that senior management spend an adequate time devoted to safety issues, with front line workers. Companies should make their own judgement on how much time their leaders should spend in the field, but best practice suggests at least one hour per week should be formally scheduled in the diaries of senior executives for this task. Middle ranking managers should have one hour per day devoted to it, and first line managers should spend at least 30% of their time in the field.

5.24   Senior management must demonstrate to their organisation in this and other ways that safety is of the highest priority and that improvements in safety will, in addition to reducing injuries and incidents, result in improved business. The "noise around performance" must be tempered to ensure it does not swamp the noise around safety. It must also be made clear to the organisation that the management of safety is the responsibility of the direct management line, and that the safety professionals are used to support, not replace, the line management. I recommend that where one is not already in place, a strategic safety management leadership team should be established in each company in the rail industry. Such a team should be led by the Chief Executive and include his or her direct reports, with support from the safety professionals. It should consider the strategic management process for safety by holding regular meetings where health and safety issues are the only issues discussed. It should be the key group in the organisation for setting goals, monitoring performance and assessing and resourcing the needs of the organisation to ensure that the long-term objectives are met. I do not wish to go so far as to be prescriptive regarding the frequency or duration of meetings of such a group, but it is unlikely to fulfil its obligations unless it meets at least bi-monthly. The outcomes from this meeting should be disseminated throughout the organisation.

Communications

5.25   It is clear from the above that a key task for management concerns communications and, specifically, communicating to all employees the clearest possible message of their safety goals and objectives. Mr Tunnicliffe described the role of leadership as "about causing people to share vision and share passions to achieve things" and said that

> "…if you can get people to share your vision at every level, then you are halfway to achieving what you are trying to achieve, because when the vision is shared, then people work at those processes and their own individual areas of leadership to achieve the same vision".

5.26   Mr Kooger said that he saw a requirement for a small number of "golden rules" that were

> "…so important that everyone could know them, was supposed to know them and was supposed to live by them".

A similar view was expressed at the seminar on the Management of Change. The industry's leadership needed to think and consult about what it wished to achieve, then draw up a mission statement and a set of values, with a clear vision of where the organisation wanted to go and how. Again, the Inquiry heard good examples of how this approach was being attempted in parts of the industry, and safety was being made a visible element of the management process. However, it was equally clear that such an approach was not universal. In my view, the industry as a whole, and the individual companies within it, should agree and widely promulgate a universal set of "golden rules" for safety which govern the behaviour of employees at all levels and at all times.

5.27   The issue of communications was also considered in the seminars on Employee Perspectives on Rail Safety and Developing an Effective Safety Culture. In the former there was a clear recognition of the importance that communications played in employees' understanding of the direction of their company, its future and how they formed part of that future. The seminar also recognised that communications were, or should be, a two-way process, and hence provided an essential feedback to management as to what was happening at the working level. Effective communications encourage employees to feel valued. Good communications, it was stressed, fostered trust and respect between management and employees. They were not about telling people what they should do; rather, they were concerned with involvement and participation.

5.28   Concern was expressed at the seminars that the commitment of senior management to safety was not felt on the ground, and that employees' concerns were not adequately relayed to senior management. Dr C A Woolfson, Director of the Faculty of Social Sciences Graduate School at the University of Glasgow and Director of the European Centre for Occupational Health, Safety and the Environment, who is an expert in human relations, put it to the Inquiry that a "hidden transcript" could exist within an industry. By this he meant that management were not kept informed about the real situation in the industry but, rather, were often told what employees thought they wished to hear. He said that in his opinion this problem was worse in non-union situations.

5.29   The seminar on Developing an Effective Safety Culture also heard the suggestion that some front line staff and middle management were confused owing to the lack of clarity from the top about company policies, while information from the front line was not reaching the top of the organisation. In some organisations it was alleged that there was almost a "military approach", with the top deciding and the rest following. Comments and concerns from the workforce did not seem to be given as much consideration as they should, and in some cases were lost. The feedback of ideas and reporting of incidents were either not happening or not effective, which again made the employees feel that they had no valid input to the system.

5.30   Mr Kooger stressed the importance of person to person communications. He likened them to a bridge and referred to a process of establishing communication highways between the top and the bottom of the organisation. He discussed the development of modern management communication techniques and how the process had moved from a situation where "the boss was the boss and he gave the orders and would you please shut up", to a modern process where the entire organisation delivered the message and ensured it was understood. It was his opinion that to foster good communications it was important, firstly, to limit the number of subjects to be discussed and then to discuss those subjects by every possible means. He went on to stress that bulletin boards and written communications were very ineffective ways of communication. He preferred a process

"...to force management out on the floor, walking the job, talking to people, listening to people".

Mr Kooger also stressed that when management were out communicating with staff it was important to recognise positive contributions and to comment on them.

5.31   One essential task for leaders of all companies is to communicate the direction for their company through a mission statement or policy. Such a statement is indeed required to be shown in the safety case. The examples shown to the Inquiry, often signed by the chief executive in person and typically of one page length, included a general statement of the company's commitment to safety and to individual safety policy objectives. Some companies took the opportunity of their annual safety plans to repeat these commitments. While these policy statements may be acceptable as far as they go, and in the context of safety cases, what was less clear to the Inquiry was how they became active mission statements playing a direct part in the daily life of all companies and their staff, and how their message was conveyed to employees as a genuine, living commitment. This concern was apparent to DuPont in their report on Safety Management in the Railway Group, which noted in regard to Railtrack Line:

"While good safety policy statements exist for Railtrack and Zones, they are not seen as working documents, which are communicated to all employees, and become part of how all employees perform their duties".

5.32   Below this top-level safety policy the requirement for clear safety rules and responsibilities is equally plain. The lack of clarity of some industry rules was one of the issues underlying the relevant accidents which were identified by ATOC in their statement of case. In particular ATOC highlighted the lack of clarity and effectiveness of long-standing practice and rules as demonstrated by the Southall crash. Entec's review of the relevant accidents found a deficiency in the clarity of contractual responsibilities for performing safety-related functions and checks. Indeed most of the relevant accidents raised issues as to the clarity and ownership of rules, contractual and other responsibilities, and accountabilities.

5.33   Thus, for example, in the case of the Watford South accident, the HSE concluded that the

"...wording of SSP 20 is imprecise and has given rise to different interpretations as to which signal should have the Speed Restriction on its approach".

After the Bexley accident the HSE concluded that it was unclear who was responsible for progressing the replacement of the longitudinal wheel timbers, and that

"...there was no clear and consistent picture for ensuring works were undertaken or identification of who was responsible for championing the work".

In the case of the Norton Junction incident there was a failure to identify and rectify a wrong-side failure of an Automatic Warning System (AWS) indication. Contractors maintained that the failure of their fault teams to identify the failure was due to a defect in the applicable Group Standard. In the case of the Southall crash, Professor

Uff concluded that there was an appalling lack of clarity in what was said in the Rule Book provision about the operation of trains without AWS.

5.34   As regards other forms of communication, the seminar on Employee Perspectives on Rail Safety discussed the process of team/safety briefings, which were seen as one of the cornerstones of developing a safety culture. The seminar preferred the term "safety meeting" since "briefing" implied a predominantly one-way communication. It was alleged that the quality and standard of such briefings or meetings varied. In some companies they were very good, but in others they had been discarded. In one case safety briefings had been subsumed into regular team briefings, with the meetings tending to concentrate on performance statistics. It is important that such safety meetings should link directly to the safety management leadership teams that I describe in para 5.24. There should be a two-way communication process between management and the workplace by this means. The seminar also noted difficulties with incorporating the meetings into the shift-working pattern. Management must take any steps required to overcome such problems.

5.35   Some examples of other methods of communication with front line staff were discussed. These included, for example, SPAD warning sheets, which gave much more detail of the problems at a particular signal. Others commented positively on videos and interactive systems that had been introduced in some companies. It was, however, also alleged at the seminar on Employee Perspectives on Rail Safety that the safety days which were instituted as a consequence of one of the recommendations of Sir Anthony Hidden's report on the Clapham crash were now on the wane, with safety and team briefings replacing them.

5.36   A confidential reporting system such as CIRAS could be said to tap into the informal structure of an organisation, and could thus be used to gauge what is actually happening and why. The fact that there was a need for such a system was seen by some to be an indictment of the communication process. Participants in the seminar considered that if there were an appropriate and just culture within an organisation rather than one of blame, then there would be no need for a confidential reporting system.   Professor Baldry, Professor of Human Resource Management at the University of Stirling, who is an expert on human resource management, welcomed the use of CIRAS but said:

> "I think the fact that CIRAS was necessary is a bit of an indictment on the approach to safety in the industry up to this time. I think that the fact that you have to have an anonymous system of reporting speaks volumes in terms of the consequences for those who reported dangerous practices pre-CIRAS".

5.37   It was clear from the seminar on Public Perceptions of Rail Safety and the evidence of Mrs Deirdre Hutton, Vice Chair of the National Consumer Council, that the industry has an equally important task to do in improving communications with the public. As I note in Chapter 9, she said that the public are not clear about who is responsible for what within the industry. Mrs Hutton also supported the view expressed at the seminar that the public's trust and confidence in the railways had declined, that they wanted better information and that their views needed to be taken into account. For example, they heard about larger-scale incidents on the railways but not about the full spectrum of safety issues. It was also felt that a regular supply of information from the industry was to be preferred to one-off responses.

Staff motivation

5.38 There can be no doubt that, if the rail industry is to reach the level of performance required, highly motivated staff at all levels will be required. Staff motivation derives from many aspects of the management process. The seminar on Employee Perspectives on Rail Safety painted a bleak picture of the current state of motivation. It was alleged that employees were frustrated as they strove to meet what were, in their opinion, conflicting priorities of performance and safety. They spoke of an increase in the blame culture and referred to a loss of comradeship and confidence. Job security was an issue, with many jobs migrating to contracting companies. Drivers' representatives spoke of intolerable pressures to run to time and to cope with the paperwork burden required to explain deviations from the timetable. They felt that professionalism was placed in question, and that they were required to run with equipment which was in a less than ideal state. It was alleged that pre-privatisation one might be disciplined for a deviation which post-privatisation would result in termination of employment.

5.39 ATOC disputed the picture of the industry painted at that seminar. They referred to the input to the Inquiry by many senior managers from all but one of the TOCs, and claimed that there was a high level of commitment and dedication shown by them. They pointed to steps which those managers had taken to improve morale, such as the formal and informal safety tours described by Mr Baker, Deputy Managing Director of Northern Spirit, and Messrs McClean and Bird, and work sharing, e.g. assisting in the buffet or riding with drivers. They alluded to the value which they attributed to union representation, as described by Mr Baker, who said:

> "The health and safety representatives are very much involved and we take a very strong view within Northern Spirit about the involvement of the union health and safety representatives. I mentioned that the control distribution of the safety case and our procedure manuals is down to junior management level. Equally, all our health and safety representatives have copies of the safety case and the manual of safety procedures. Recently, we have sent our company safety committee on the same training course that the board members are required to go on in strategic safety management, and we are absolutely committed to getting that wide input".

5.40 Mr B R Burdsall, Managing Director of Midland Main Line and Mr Kooger also stressed that non-union employees were consulted as well. There are thus differing views of the state of morale in the industry. The actual position may well be somewhere between the positions which were described, but there is general agreement that morale can and must improve. The steps described in this chapter to improve the safety culture should bring a concurrent improvement in morale as people feel the effects of success. There was general agreement that reward and recognition programmes should be linked to success to improve both performance and morale, and several companies told the Inquiry how bonus systems were linked to safety performance. Mr Kooger spoke of the need to give praise whenever it was due, and to reinforce positive achievements.

Toleration of unsafe acts and the "blame culture"

5.41 There is a need for the industry to develop further as a learning organisation. By this I mean in particular that the industry should learn the lessons from previous accidents, near misses and the analysis of information regarding the non-compliant behaviour of people and systems; the analysis of behaviour leading to unsafe acts; and incidents in other related industries. It is important also that the lessons learned are communicated to all parts of the industry, and actions are taken to prevent a repeat of these incidents.

5.42 In this context, the Inquiry had the benefit of a joint statement by experts on the use and application of risk assessment in the rail industry, which was spoken to in evidence by Mr R Sylvester-Evans, Safety Consultant, who chaired the meeting of experts. The Inquiry is grateful to those whose work contributed to the joint statement and its annexes. Since this is of value and should be made more widely available, it is set out in Appendix 7. Mr Sylvester-Evans, himself an expert in the subject of risk management, was questioned about the use of, and interactions between, risk management, cost benefit analysis and the ALARP (as low as reasonably practicable) principle. He explained that cost benefit analysis was a tool which was employed to assist in the determination of whether there was a financial justification for a course of action. However, in his opinion, it was not a stand-alone management decision process. He agreed with Counsel for the Collins Passengers' Group that the ALARP process could be summarised as follows:

(i) identify significant hazards which lead to the element of risk;

(ii) list all practical risk reduction measures for the identified risk;

(iii) assess whether risk reduction is possible, using, for example, cost benefit analysis; and

(iv) repeat the process to ensure that all practicable and reasonable risk reduction measures are identified and implemented.

He went on to illustrate how the use of risk assessment can lead to continuous improvement, since management are required to assess their operation continually as systems and technology improve, and to assess their compliance with the ALARP principle. In his opinion this was a more exacting process than a standards-based assessment, for the latter, once the safety situation had been assessed, would be unlikely to be revisited unless the standard was changed. Counsel for the bereaved and injured represented by the Southall and Ladbroke Grove Solicitors' Group supported the need for the industry to develop further the use of risk assessment, and referred to Part 1 of the Inquiry which they contended highlighted significant failures in its use, namely the failure to carry out risk assessment on the revised Paddington layout, and the use of risk assessment to justify inaction. I commend the use of risk assessment in the rail industry and I agree with the joint statement of experts that

"...future railway risk assessments should take more account of the complex interactions between the trains and the infrastructure".

5.43   Under cross-examination, Mr Waite agreed that the training of management to ensure that staff recognised unsafe acts should be added to the list of safety management deficiencies noted in his review of the relevant accidents as safety management deficiencies.

5.44   The existence of multiple SPADs at a particular signal is a clear example of what is effectively an unsafe condition. Whatever the cause of the SPADs the repeated occurrence indicates an underlying problem, which, if not identified and solved, represents an acceptance of that unsafe condition. The Watford South accident was the occasion of the fifth SPAD at the signal concerned within two and a half years. The Ladbroke Grove crash occurred with the ninth SPAD in just over five years.

5.45   The Southall crash provides examples of the acceptance by management of sloppy practices and unsafe conditions. Professor Uff observed that there were three different and separate specifications covering the testing of the AWS. This led to uncertainty as to what was required. There was a failure by Great Western Trains to take a train out of service when the AWS was isolated, and an absence of any appropriate procedure for this situation.

5.46   In my report on Part 1 of the Inquiry, at paras 9.57-9.61 and para 11.32, I discussed the question of a "no blame culture" in the industry and I now return to this subject. In the seminar on Developing an Effective Safety Culture, many participants pointed to the need to eliminate the barriers to the development of an effective learning process. In the Inquiry, the Joint Rail Unions restated their concern that the "blame culture" inhibited the reporting, and prevented a full examination, of incidents. It was suggested that the key characteristics of a healthy culture were a reporting culture (of which CIRAS is a part); a just culture (e.g. looking for root causes); a flexible culture; and a learning culture.

5.47   In their closing submissions to the Inquiry the Joint Rail Unions mentioned the CAA's Mandatory Occurrence Reporting system which is designed to ensure that all accidents are reported and acted upon. They drew particular attention to the fact that incidents and occurrences could be reported anonymously and that confidentiality is respected throughout the investigation. However, I consider that the CIRAS system provides adequate arrangements for reporting within the railways.

5.48   The Unions also advocated in their final submission that accredited and trained safety representatives should be empowered to issue provisional improvement notices, subject to appeal by the employers. This was supported in evidence by Mr V G Hince, Senior Assistant General Secretary of the RMT, who stated that this was reflected in the TUC's response to a paper by the HSC regarding greater involvement of trade union representatives in the work place. There is obvious merit in all employees having the duty to consider the safety aspects of their assignments, bringing to management's attention matters of concern, and in the final analysis enlisting the support of other agencies if they feel there has been an inadequate response. However, it seems to me to be premature to introduce provision for provisional improvement notices. A significant degree of training would be required before such a step was taken.

5.49   It was the Unions' view that further progress needed to be made in these areas. The Railway Group Safety Plan for 2000/01 also recognised this need and said that

> "...a thorough and wide ranging review of safety systems which can be expected to further reduce the levels of risk arising from human error is required.
>
> The need for the Railway Group to address the portion of risk attributable to unsafe acts by persons other than the workforce or passengers (62%), and failures in safety management systems and workforce accidents (22%) is essential if significant progress is to be made in achieving the long-term goal".

The plan went on to point out that the industry should look to the underlying hazards and root causes of incidents, and learn from these events. The DuPont report on Safety Management in the Railway Group also stressed

> "The culture that near misses are an opportunity to learn and improve needs to be strongly developed to overcome the current tendency to attribute blame. This can be done through incident investigation training for management and supervisors, the pro-active free sharing of information from incidents, and a relentless drive to implement corrective actions".

5.50   Speaking to this report in his evidence Mr Kooger again stressed the need to work on the management systems and processes that caused unsafe acts. He pointed out that it was necessary to attack all unsafe acts and not to try to pick out those which might have particularly serious consequences, because in practice, any unsafe act can have a serious consequence. In this context Mr Kooger criticised the Railway Group Safety Plan because in his belief it did not focus clearly enough on unsafe acts. He also considered that the goals set for improvement in the plan were not demanding enough. It was his experience that a focused attack on unsafe acts and the lessons from them could result in a 30% year on year improvement in all aspects of the safety performance. He compared this to the goal of 50% improvement for the ten years of the period of the Railway Group Safety Plan.

Continuous learning

5.51   One particular criticism of the "blame culture" was said to be that it inhibited the industry from the proper investigation of accidents and incidents, and the open sharing of information. Thus Mr Waite noted that while investigations into accidents often claimed to find underlying causes, this was not always the case: there was a reluctance to search for industry-wide or system problems. Instead, investigations concentrated on local faults. He said:

> "...the industry-wide blame culture and fragmentation has hindered the efforts to identify the fundamental problems. The immediate cause may only be a symptom of the underlying cause".

In the words of the DuPont report, there was a tendency to look for "the guilty party rather than the act and the reasons behind it". The blame culture was also said to inhibit staff from reporting non-serious accidents. The significance of this weakness was clear from the emphasis which a number of witnesses laid on the importance, within a successful safety culture, of continuously learning the lessons of accidents and incidents in order to prevent them from re-occurring, and to having clear and consistent processes to enable that to happen. An organisation which learned from accidents, incidents and failures of management would be more likely to implement change than an organisation which was "in denial". The benefit of accidents and incidents is that they provide an opportunity for a company or an industry to learn from its mistakes. This is as true of near misses and other minor incidents as it is of major accidents. They should all be treated as demonstrations of the failure of the safety net. They require to be investigated with the aim of rapidly identifying the unsafe act which caused them and feeding this back for correction at source.

5.52 The Joint Rail Unions therefore called in their final submissions for the organisational changes required to produce

> "...a system that enables the full lessons to be learnt from every accident and near miss".

However, even apart from the problems of the blame culture, it is clear that good practice in the field of investigation has a long way to go for the industry as a whole. Thus, for example, Ms A E Forster, Operations and Safety Director for First Great Western (FGW), spoke of the "very bureaucratic process" which the industry had developed for investigating accidents and incidents. It led to delays in getting recommendations and reports out to the industry. ATOC stated that there was no co-ordinated process for collating recommendations which dealt with the same areas, so as to ensure consistency and promote monitoring and action. Mr Bird, who spoke to ATOC's statement of case, described the "plethora of recommendations". Their status was unclear. There was no prioritisation. There was the risk of multiple solutions. The Entec report on the relevant accidents criticised the industry both for poor follow-up of recommendations from accident inquiries and the lack of reliable procedures for ensuring that such recommendations were carried out. As examples Mr Waite cited recommendations from the Inquiry into the accident at Newton Junction. One recommendation, which called for risk analysis of proposed schemes involving single track working, was considered to be complete, but nine years later the issue of risk analysis for high risk track layouts was still being discussed. Another recommendation related to instructions and training for the prompt and effective use of track-to-train radio systems in an emergency. The response was that technology would continue to improve (implying that no particular action was necessary), but the issue was closed with no evidence that a satisfactory stage had been reached. An internal HSE memo from Mr R Andrews to Mr Coleman written on 26 October 1999 highlighted a growing concern at the lack of progress on the follow-up of outstanding recommendations arising from the accident at Watford South. The HSE's report "The Management of Safety in Railtrack" noted a number of weaknesses in this area, including weak and variable monitoring of compliance and poor pursuit of the underlying causes of accidents and incidents. The DuPont report "Safety Management in the Railway Group" spoke of the danger of management starting a programme of corrective action with enthusiasm, only to allow it to melt away as other issues arose.

5.53   This picture is clearly consistent with what emerged in Part 1 of the Inquiry, including, for example, failures to learn from previous SPADs, to carry through inquiry findings, and to follow proper procedures for the conduct of inquiries and the implementation of their recommendations.

5.54   At the same time others felt that the industry's performance was improving in this area, not least as a result of the Ladbroke Grove and Hatfield crashes. The HSE's guidance on the Railways (Safety Case) Regulations 2000 indicates clearly that railway operators are required to investigate all accidents and incidents which could endanger people, and to co-operate, as required by Regulation 11, where such an accident or incident involves more than one operator. However there was no disagreement that more needed to be done to follow up incidents, streamline processes, improve implementation and share information.

5.55   The situation regarding safety auditing appeared to be similar in a number of respects to that regarding incidents and inquiries. It is clear that there are well established processes and systems for safety audit, but it is also clear that they are not working as effectively as they should, or producing the best results. A learning industry will view auditing as a help rather than a hindrance. It will look to the auditors to tell it as much as possible about the subjects and areas being audited, and it will want to assist them with their diagnosis through a co-operative approach on its own part. Some described "death by audit", where process too often seemed to take precedence over learning lessons and taking action. The DuPont report found that much time and effort appeared to be spent on policing, fault-finding and excellence in documentation. There was a great deal of repetitive auditing, with no associated drive to implement corrective actions. Again, the evidence heard in Part 1 of the Inquiry reinforced this picture.

5.56   A further aspect of the industry's capacity for learning related to its use of data and analytical tools. A number of concerns were voiced about weaknesses in this area. Thus ATOC stated in their closing submissions:

> "The industry has a wealth of knowledge, research and data but it is not always known or locatable"

and:

> "SMIS contains a wealth of data. However, to become a more readily usable tool it needs to be far easier to access and use".

The Collins Passengers' Group stated in their closing submissions:

> "The safety regulator should take more active steps to ensure that risk assessments are properly understood and carried out by duty holders".

Ms S A Brearley, Controller of Safety Strategy and Planning in the S&SD, stated in her witness statement:

> "At industry workshops in November 1999, only two participants out of around 80 acknowledged having used the cost benefit decision framework".

She also stated:

> "In my view, the industry's tardy adoption of risk assessment methods is now being addressed. Within S&SD there is evidence of development of more sophisticated risk methods and risk thinking. Other parts of the industry are adopting the methods with various degrees of enthusiasm. The way is clear for improved use of risk information, but a significant hill remains to be scaled".

Training and competency

5.57   It is clear that a learning organisation will set greater store on training and the development of competency. The need to develop the amount and quality of training within the rail industry was another recurring theme throughout the Inquiry. Beyond the problems which were identified as flowing from privatisation, and which are described in Chapter 4, it is clear that there is need for progress in this area. The seminar on Employee Perspectives on Rail Safety revealed mixed views as to whether training was adequate. The majority view was that initial training was good, but there were concerns in regard to refresher training, which many people thought was directed towards recent recruits to the industry and did not adequately address the needs of those with longer service. It was suggested that pre-privatisation training standards of performance were set for employees at various stages of their development but that this had now stopped. It was also pointed out that companies had different cultures which affected the view and perception of training. However there was unanimous agreement as to its importance.

5.58   Mr Bird said:

> "I believe training is the cornerstone to any improvement".

Dr Scott described how in Eurotunnel training was delivered at all levels in the company, from leadership right through to the direct workplace, and mentioned the need to see that the training was delivered in a co-ordinated manner. Mr Tunnicliffe explained how in his opinion good training helped employees understand the overall picture of how the company worked.

5.59   ATOC identified training as one of the underlying, and recurring, themes of the relevant accidents. The accidents at Newton Junction, Bexley, Southall and Ladbroke Grove all raised issues about the training of drivers, signalling staff or both. The Watford South accident report required the TOC in question to provide a programme of action to reduce substantially the number of incident-prone drivers. Following the Southall accident Professor Uff made recommendations with regard to the training and competence of controllers and supervisors in transmitting, receiving, recording and acting upon safety-related messages. Following the Newton Abbott derailment, which was due to a failed axle, there was a recommendation re-emphasising the need for correct testing, which resulted in a programme of retraining of non-destructive testing operators. Arising from the Norton Junction wrong-side failure of the AWS ramp, recommendations were made by the S&SD for the improvement of the knowledge and training of contractor staff involved in fault finding.

5.60   Professor R A Smith, Head of the Department of Mechanical Engineering at Imperial College and Chairman of the Advanced Railway Research Centre, described in his evidence the Japanese approach to training. This was also a subject covered in detail in the seminar on The Japanese Model of Rail Safety. In Japan continuous training is seen to be part of the job, and its use fosters pride and status. Training incorporates an integrated approach whereby all grades come together for some aspects of the training. Full-scale facilities are provided for training in avoiding hazard or disruption on the network. The Inquiry heard how, for example, drivers were questioned on a day to day basis about specific subjects, such as speed limits and route knowledge. The extensive use of on-train monitors was also noted. The Japanese railway companies make extensive use of technology with a considerable concentration on simulators. These are programmed to help drivers learn about train failures and emergencies. Drivers are expected to demonstrate their capabilities to their supervisors and peers in a two-monthly simulator test.

5.61   Many positive aspects of training in the British rail industry were described. Mr Muttram told the Inquiry, for example, how Railtrack had arranged for their Board members to be trained in observing safety practices in the workplace. He described the instruction which they had been given on the importance of a strategic approach to training. Mr C Carr, Technical Director of Amey Rail, described the work of the RITC and how they had recently been accredited to issue National Vocational Qualifications (NVQs). He told the Inquiry that while the number of employees taking these qualifications was as yet low it would increase.

5.62   Several witnesses, however, spoke of the need to increase training effort. Mr Bird agreed with the suggestion in the DuPont report that driver standards managers should review training in root cause analysis, and Professor J B Davies, Professor of Psychology at the University of Strathclyde, stressed the need for more safety training for middle level managers. Mr J Knapp, General Secretary of the RMT, Mr R Rosser, General Secretary of the TSSA and Mr M D Rix, General Secretary of ASLEF, speaking together for their Unions, gave their support for more training for their members and stressed their willingness to co-operate with it. They made the point, however, that in their opinion, training should be scheduled into the work roster. Mr Knapp pointed to the need to train more people to replace those who left at the time of privatisation. He expressly referred to the need to provide paid release for the training of safety representatives.

5.63   I strongly endorse the concept that the industry should pay particular attention to the continuing training needs of all levels of staff.

"Interdependency"

5.64   I now return to a consideration of the culture which I outlined in para 5.9. However, I suggest that the industry should develop a behaviour which leads to all members of an organisation, and all collections of organisations, working together for the common good of a safe railway.

5.65   Counsel for the Inquiry expressed the development of this culture as

"…a progressive movement from a situation of dependency, where management makes the rules and tells employees what to do, to a situation where individuals can contribute ideas and effort, while complying with the rules and procedures, through to a position where there is a committed, dedicated team approach, with a high degree of interdependency between teams and across company boundaries".

5.66   The evidence discussed in Chapter 4 specifically, and more generally elsewhere in this report, suggests that the rail industry has not yet reached this desirable state of interdependency. I consider that the industry and its component companies should develop their management processes to ensure that the culture is driven in this direction.

5.67   In para 5.17 I point to the need for a rail industry body to provide leadership. It should enable safety matters relevant to the whole industry to be taken forward, with the various companies giving support to one another by sharing information and resources as appropriate. For this to be effective it will be necessary for the companies to be less competitive in their relationships, looking towards the long-term health of the industry rather than to the short-term advantage of the individual company. The approaches suggested elsewhere in this report and the report for Part 1 of the Inquiry, for example, joint training of drivers/signallers, central licensing and "golden rules" across the industry, should play a major role in the cultural shift which is required. Within individual companies the concept of team working at all levels should develop an interdependent culture, within which people will help one another and, if needs be, subjugate their own interests to those of the business, and especially to the safety of everyone on the railways.

Conclusions

5.68   While there are encouraging signs of good practice, performance across the industry as a whole is at best patchy. The way forward is clear. The industry needs to take all necessary steps to set high safety standards through clear leadership; good two-way communications; a relentless pursuit of excellence of operations through the identification and adoption of best practice, learning processes, training and the involvement of all employees; a new focus on the real concerns and interests of customers; and a new ethos of co-operation across the industry.

Chapter 6
Railway Group Standards

Introduction

6.1   The purpose of this chapter is to set out a number of aspects of the development and use of Railway Group Standards. It should be read along with the outline of this subject which was given in paras 3.49-3.55. The question of the body or bodies which should be responsible for the setting of Group Standards and auditing compliance with them is discussed in Chapters 8-10. In considering the subject of Group Standards the Inquiry had the benefit of a report by Entec UK, which was spoken to by Mr W A T Alder, Principal Consultant for Entec UK.

6.2   This chapter deals with the following subjects:

- the Group Standards (paras 6.3-6.5);
- content and significance (paras 6.6-6.15);
- the production of Group Standards (paras 6.16-6.18); and
- auditing (paras 6.19-6.20).

The Group Standards

6.3   The present Railway Group Standards, which include the Rule Book for train drivers and other railway staff, have their origin in the standards, primarily of a technical and operational nature, which were formulated and applied by British Rail. They were used as a mechanism for controlling the construction, maintenance and operation of the railways for which British Rail were responsible. Their objective was the safety of the system and its trains, and hence the safety of passengers and railway staff.

6.4   Railtrack assumed responsibility for these standards in so far as they related to the safety of the system and safe interworking. Railway Group Standards may be categorised as comprising technical (engineering) requirements, arrangements governing the relationship between trains and the infrastructure, operating requirements and procedural matters. Railtrack undertook the process of reviewing them in order to rationalise and simplify the very large number which had been taken over. Through this process Railway Group Standards were reduced to just over 500. It is evident that this work has still some distance to go. Mr Coleman applauded what had been done so far. To some extent the existing Group Standards, particularly those relating to technical requirements for equipment, will require in due course to be replaced by European standards. This is not likely to produce particular difficulties, but may well take a significant time to achieve.

6.5   As I have already noted in para 3.55, Railway Group Standards do not apply to matters which are entirely within the control of a TOC and do not affect the safety of the staff or passengers of any other TOC or the general public. In this category are about 220 standards related to the design of train interiors. It is expected that these will be reduced to about 30-40.

Content and significance

6.6　As was pointed out in the interim report of the HSC "Review of Arrangements for Standard Setting and Application on the Main Railway Network", railways are a tightly organised transport mode. Accordingly their safe functioning is critically dependent on

> "...the continuing satisfactory condition of vehicles, infrastructure and personnel and the interfaces between them. Design and performance need to be closely matched between the train, structure and signalling" (para 22).

6.7　It is clear that Railway Group Standards represent a key element in the effective control of risk.

6.8　Railtrack treat and have treated Railway Group Standards as part of the conditions on which train operators are permitted to operate their vehicles on Railtrack's network. Railtrack's position is that if Railway Group Standards did not exist, they would have to impose their own conditions of access to ensure that their statutory duties were met. Mr R I Muttram, Director of the S&SD, said that legally Group Standards were Railtrack's standards. However, the objective of including people from across the industry on the board of Railway Safety was

> "...to give greater assurance that these standards in the way they are constructed are fair and equitable to all".

6.9　The formulation of a Group Standard may be based on detailed technical and scientific evaluation. The fact that, when this work has been done, it does not have to be repeated by individual operators means that there is a common benefit to those to whom the Group Standard applies. A Railway Group Standard may be specifically concerned with the cause of a past accident, with a view to preventing its repetition. However, where this has been done, it is, as Entec observed, a reactive rather than a proactive approach to safety. It should not be the only way of achieving safety improvement.

6.10　Group Standards range from the highly prescriptive to those that set out goals. The latter give freedom as to how their objectives are to be met, whereas the former are used where goal-setting is not adequate, for example, where consistency of action is essential to deal with the complexity of interfaces. It is, of course, well recognised that for safety improvement it is essential for managers and the workforce to understand the hazards and risks which their activities involve. Standards which are prescriptive provide less opportunity for their involvement. They can lead to a culture of mere "compliance". Thus standards which set minimum acceptable requirements may be implemented without regard to whether risks have been reduced as low as reasonably practicable. Prescriptive standards can also prevent innovation and improvement in designs and working methods.

6.11   At the Inquiry differing views were expressed as to whether there should be a move to greater prescription. No particularly compelling argument was advanced for this. On balance the present arrangements, which provide for both prescriptive and goal-setting Group Standards, according to the nature of the risks with which they are concerned, appear to be satisfactory. The Entec report drew attention to the fact that the HSE report "The Management of Safety in Railtrack" (at para 46) pointed out the need for a documented strategic policy. The Entec report pointed out that this could be used in order to give guidance as to the balance which should be observed between prescriptive standards on the one hand and goal-setting standards on the other. Some parties had reservations about the mix of prescriptive and goal-setting standards, maintaining that they should be clearly separated. While such a separation may be ideal, it is sufficient, in my view, that it should be one of the objectives in the ongoing rationalisation of Group Standards. I should add that it is desirable that for a single operating procedure there should be a single Group Standard (cf Recommendation 11 of Professor Uff in his report on the Southall crash).

6.12   Mr Muttram gave evidence that a system of Group Standards was essential and that it should be

> "…based on the principle of continuous improvement that underpins TQM (Total Quality Management)".

Some witnesses expressed the view that there was a lack of a common benchmark or target against which an existing or proposed Group Standard could be assessed. Para 6.3.1 of the Railway Group Standards Code states:

> "There is an overriding Safety Criterion which requires the outcome of all decisions demonstrably to contribute to levels of risk that are 'as low as reasonably practicable'".

However, one of the principal findings of the HSE in their report "The Management of Safety in Railtrack" was that it was not transparent how the process of standard setting aimed to deliver this (para 4 of Appendix 1). Entec commented that it was possible that standards merely set out the bare minimum for achieving safety "rather than providing activities that are further into the safe region".

6.13   Entec noted that both the HSE in their report "The Management of Safety in Railtrack", and DuPont Safety Resources in their report "Safety Management in the Railway Group", considered that the Railway Group Standards system should be used to provide guidance on safety management systems. I consider that there is merit in Entec's observations that great care should be taken in the setting of standards for this purpose. They stated:

> "It is important that an organisation's safety management system is designed as the best option to suit their way of operating. Such a system is likely to have many features that are unique and suit that organisation. In addition, to achieve ownership of safety at all levels within an organisation, the development of the system needs to be a learning process. It would not be appropriate to provide detailed standards that tell organisations the details of what their safety management systems should be".

6.14   All duty holders have standards of their own covering matters within their own responsibilities. Railtrack set Line Standards for the functioning of their network and stipulate that train operators and contractors require to comply with them. While Group Standards set out the policy and what is to be achieved, Railtrack Line Standards are directed to technical requirements and the means to be used to meet the stated end. Mr C E Perry, Group Managing Director of AEA Technology Rail and Chairman of the RISSC, observed that Line Standards tended to be more prescriptive than Group Standards. The consultation regime in connection with them was less well defined. The boundary between Group Standards and Line Standards was fairly clear but it was not well understood in the rail industry.

6.15   Mr R H McClean, Production Director of GNER, said that unfortunately a number of standards relating to the interface between trains and track, trains and signalling and trains and overhead lines had "migrated" from Group Standards to Line Standards:

> "We, therefore, as train operators, do not have visibility of changes that take place there or the processes that validate or justify those changes".

The production of Group Standards

6.16   The production of a Group Standard follows a considerable amount of consultation that is carried out both within standards subject committees and with other interested parties, in accordance with the Railway Group Standards Code. The principal means of consultation are the subject committees and ad hoc groups. Representation on the subject committees includes the ROSCOs and maintenance contractors. The rail trade unions are among those consulted. It is clear that such consultation with the industry is necessary so that as far as is possible there is agreement in regard to a new or revised standard prior to its being put into place. It is equally clear that, whatever body is to be responsible in future for the setting of Group Standards, the need for the thorough consultation of all interested parties will remain. However, the process has its disadvantages. In their report "Safety Management in The Railway Group", DuPont Safety Resources stated (at page 31):

> "The existing procedure for obtaining approval for changes to Railway Group Standards is cumbersome and even under favourable circumstances may take 6 to 12 months. This is dissuasive to individuals with innovative ideas and unduly extends the timely implementation of cost saving and safety enhancing suggestions. A rapid response procedure exists for use in exceptional circumstances. Unfortunately these delays have resulted in successful efforts to circumvent or even avoid the management of change procedure".

The HSE are always consulted on a proposal for a new or revised Group Standard, and particularly examine the proposal in order to ensure that risks are reduced as low as reasonably practicable.

6.17   One particular group which is not involved is passenger representatives. There is a view in some quarters that they should be involved. Others consider that the task of drafting and approving standards of a highly technical nature should be carried out by those who have the technical ability to do so, and have a part to play in the application of, and compliance with, such standards. In my view, if passenger groups are to be involved – and it would seem to be in the interests of public confidence that they should be – this should be at the higher controlling level of the whole process. In this way they can make a contribution to policy issues in regard to standards and their application without being involved in the detail. It should be noted that the Rail Passengers' Council is already represented on the SAB.

6.18   It should also be noted that, as was pointed out in the report "The Management of Safety by Railtrack" at para 228, while individual Group Standards are reviewed, there is no systematic review of the standard setting process to assess whether it is effective in achieving its overall aim of safe interworking. I agree that this is an area for improvement.

Auditing

6.19   At present Railtrack monitor compliance with Group Standards as part of their auditing of compliance with safety cases. Evidence was given that 38 Group Standards were subject to checking by auditors as part of this process. The auditing of Group Standards was an area where it was considered that present arrangements require to be strengthened. This was seen by many, however, as a sample assurance process rather than the full control process which was said to be needed. However, Mr Perry expressed the view that the present process was adequate, with the HSE overseeing the arrangements on behalf of the public.

6.20   Compliance with Group Standards also falls within the scope of auditing by the HSE. Failure to adhere to a Group Standard is treated by the HSE as failure to comply with a safety case. This may lead to enforcement action. In their statement of case the HSE gave as an example the action taken because of the failure of Great Western Trains to ensure that the train which was later involved in the Southall crash was taken out of service when its AWS was found to be faulty.

Chapter 7
Safety cases, accreditation and licensing

Introduction

7.1 This chapter is primarily concerned with the application of the safety case to railways in Great Britain, and with the working of the current RSC regime. It is not concerned with the functions of Railtrack or Railway Safety in regard to the regime, or with the question whether there should be changes in the way in which it is administered. A number of parties took the opportunity of advocating a system for the accreditation of companies such as maintenance contractors which supply products or services for the railways. There was also discussion on the licensing of individuals whose work involved significant risk. These matters are also included.

7.2 The Inquiry heard the evidence of a number of witnesses from the rail industry, principally Mr R I Muttram, Director of the S&SD (who has since become Chief Executive of Railway Safety); Mr S K Baker, Deputy Managing Director of Northern Spirit; Mr C Carr, Technical Director of Amey Rail; Mr G C Eccles, Director of Stagecoach Holdings; and Mr H T Abbott, Managing Director of Angel Trains. In addition the Inquiry heard the evidence of Mr V P Coleman, Chief Inspector of Railways for the HMRI; Mr M H Brown, Assistant Chief Inspector of Railways for the HMRI; and Professor A W Evans, Professor of Transport Safety at the Centre for Transport Studies at University College, London. Mr P J Waite, Technical Director of Entec UK, spoke to a report on the safety case regime which had been commissioned from Entec for the assistance of the Inquiry. The report included observations on a representative number of safety cases.

7.3 The subjects covered by this chapter are:

- the appropriateness of the safety case for Great Britain's railways (paras 7.4-7.9);
- the control of risk at sites and interfaces (paras 7.10-7.15);
- the working of the safety case regime (paras 7.16-7.46); and
- an extended safety case regime, accreditation and licensing? (paras 7.47-7.75).

The appropriateness of the safety case for Great Britain's railways

7.4 In their report "Ensuring Safety on Britain's Railways" in January 1993 the HSC recommended, in the light of impending legislation for the privatisation of the rail industry, that the future safety regime should be based on the safety case. It was clear that the control of railway operations, on which responsibility for safety depended, would be divided between many different organisations. The HSC stated:

"This will generate a need to define the extent of the responsibilities of each party and ensure effective management of safety (including emergency planning) particularly at the numerous interfaces between parties".

The safety case was and is seen as providing, in the disaggregated state of the rail industry, an appropriate means of managing safety on the one hand and, on the other hand, providing an adequate assurance of safety for independent scrutiny. Under the Safety Case Regulations an accepted safety case requires to be held by an "infrastructure controller", such as Railtrack, and by those who operate trains or stations. Infrastructure controllers and the latter operators are collectively referred to in the regulations as "railway operators".

7.5   In this Inquiry none of the parties expressed any reservation about the principle of applying the safety case to the rail industry. Some of them drew attention to areas in which there were deficiencies in the working of the system. However, Mr P D T O'Connor, Consultant Engineer, suggested that, while such a regime was appropriate for the introduction of new risks associated with new systems, it was not appropriate for ongoing day to day operations where the risks of existing systems were already known and understood. On the other hand, as Mr Muttram pointed out, the operation of the railways involves the risk of accidents which may have very serious consequences from time to time. Mr O'Connor advocated, in place of the safety case regime, a combination of prescriptive rules and strong and competent management. The need for the latter is not in question, but reliance on the former has distinct limitations. As Mr Waite pointed out, a more prescriptive regime – say, by an extension to the scope of Group Standards – would require a large number of rules. It would still be necessary to show that their observance would be adequate to ensure safety. It would not remove the need to show that adequate procedures were in fact followed. Mr O'Connor's proposal also carries with it the unattractive implication of the transfer of responsibility from the operator, who is in control of operations, to the rule-maker, who is not. In any event prescriptive rules do not deal with changing situations, deviations from the normal and the need for continuous improvement of safety. Quite apart from these points, which I accept, the alternative suggested by Mr O'Connor does not seem to me to address the variation between the activities, locations and risks involved in the operations of particular operators; or the need to secure compatibility at the interfaces between them, which he accepted were "uniquely complex".

7.6   Mr O'Connor and Major C B Holden, a Transport Safety Consultant and formerly an Inspector of Railways, advanced a number of other criticisms, including that operators too often relied on experts for the writing of their safety cases, which moreover tended to be written in a manner driven primarily by the need to comply with the requirements of the regulations. They failed to think constructively about important issues of safety. However, these and other criticisms did not strike at the root of the regime, but rather provided examples of a poor approach to the preparation and use of safety cases.

7.7   I should add that Major Holden also asserted that safety cases tended to detract from management's responsibility for safety. However, the purpose of the goal-setting, as opposed to prescriptive, legislation relating to safety cases is to provide a framework within which management can exercise their responsibility for safety more effectively than under a highly prescriptive regime.

7.8 Major Holden suggested that an alternative to the safety case could be the holding of an annual independent audit, against a standard protocol, which looked not only at the documentation relating to safety management systems but also at what actually happened in practice. This would be following an initial validation by an independent auditor. While it is plain that effective auditing is essential, this does not seem to me to address fully the need for a framework required by legislation, within which the arrangements and procedures for the management of safety can be demonstrated and exercised in a consistent and effective manner.

7.9 In my view the application of the safety case to Great Britain's railways is an appropriate means of achieving that objective.

The control of risk at sites and interfaces

7.10 One of the matters which are critical to safety in the disaggregated state of the rail industry is the control of risk where there are interactions between the operations of different TOCs and between them and Railtrack. As Mr Muttram pointed out, the complexity of the rail system underlines the need for the safety case, in which the duty holder sets out its approach to safety, and the safety management systems which it is undertaking to apply.

7.11 In the evidence there was some discussion as to whether the safety case regime should be further developed, in the form of:

(i)   safety cases for specific sites; and

(ii)   joint safety cases.

7.12 In the offshore oil and gas industry a safety case is required for each specific site, including where different organisations combine their operations on the same site. In the rail industry, on the other hand, there is not a series of separate sites, but a single complex network in which a number of organisations are expected to work in conjunction with each other.

7.13 Mr Brown suggested that there were a number of places at which there was a need for some form of joint control of risk. There needed to be a clear understanding of the balance of control. He believed that at the points of overlap the joint construction of a safety case was necessary. He saw no difficulty in a safety case including an additional section dealing with joint operations. It did not appear to require legislation. It would be covered by the HSE's guidance. The report by Entec stated that the interfaces between different rail operators were not treated thoroughly in safety cases. Mr Waite suggested a matrix of safety cases, with each organisation producing a company-specific case and joint interaction safety cases for each interface. As an alternative he suggested that there should be:

"…joint arrangements and at least joint agreement of responsibilities and risks across the interface".

This could be based on compliance with the duty of co-operation now contained in Regulation 11. In this context the evidence of Mr D Tunnicliffe, formerly Chief Executive of London Transport and Chairman of LUL, is of some interest. He described how part of the safety case of LUL, who run the trains and stations on the London Underground, described the interfaces between them and the three companies which manage and maintain the infrastructure. Part of the safety cases of the infrastructure companies explained how they managed the interfaces. The Board of LUL treated the safety cases of the infrastructure companies as part of the LUL safety case: "They would have to fit together and serve each other".

7.14   At the same time I note that at present there are safety plans in certain places for particular railway installations. ATOC submitted that common safety plans for locations such as stations or depots could be prepared by the party primarily responsible, with input from (and possibly joint responsibility on the part of) other parties using such facilities. ATOC accepted that there was undoubtedly scope for further progress in the management of interfaces, but pointed out the difficulties of delineating them. To take account of every interface on the railways would involve many thousands of permutations. Each day on the network there were over 18,500 services, travelling over some 15,500 route miles and calling at 2,500 stations. ATOC suggested that any joint safety cases or safety statements might best operate at Zone level, although this would need further thought, on account of the number of permutations. This would help to eliminate inconsistencies.

7.15   The evidence which I have heard on this subject satisfies me that further work needs to be done on refining the definition of the responsibilities for the control of risk at specific sites which are shared between different railway operators and at the interfaces between them across the network. I am not persuaded that this calls for the production of site-specific safety cases or for extending existing safety cases to deal with the interfaces. That appears to me to be cumbersome and excessive. In my view the better approach is to ensure that the necessary arrangements and agreements are made at a subordinate level below the safety cases, and that the safety case should be used as a means of providing an index of such arrangements or agreements which will allow operators and the safety regulator to identify them readily. This can be treated as part of what is required to meet para 12 of the First Schedule to the 2000 Regulations which is concerned with the arrangements which the duty holder has established to enable it to comply with the duty of co-operation under Regulation 11. As was submitted by the HSE, it is on any view essential that a company which is preparing a safety case should consult with those companies with which it shares sites and interfaces.

The working of the safety case regime

7.16   The Inquiry heard evidence as to the manner in which the RSC regime is intended to operate, and as to the respects in which it is or is not doing so. In this section of the chapter I will set out the salient matters which emerged. In the course of his evidence Mr Waite described the purposes of a safety case as follows:

"The principal purpose, I believe, is as a tool, a route map, a record of commitments for management to set out how they organise their operation to work safely. It also follows from that and it is sometimes combined with it that it is also a document that the regulators can use to check the company's, the operation's compliance and wider than compliance coverage of all matters that might affect safety. Thirdly, it gives confidence to the regulators, and perhaps through the regulators to the members of the public, and also to the management of the company that they have adequate controls to manage their operations safely".

Mr Waite also described a safety case as being "the index page to the more detailed arrangements". Mr Brown said that its value lay in its being

"...a vehicle for communicating senior management's vision and strategy through the organisation and beyond".

Professor Evans said that he believed that safety cases were essential, being a "focal point for safety management".

7.17   Mr Tunnicliffe considered that the safety case had added more rigour and structure to LUL's safety management systems. Professor Evans indicated that there appeared to be a general consensus in the industry that the regime had created a much more systematic focus on safety than before. On the other hand the requirements of the regime are regarded by some as cumbersome.

The content of safety cases

7.18   The Inquiry heard evidence from a number of witnesses about the poor quality of certain safety cases, especially the earliest which had been produced. Professor Evans found that, despite the explicit requirements of the 1994 Regulations, a third of safety cases which he had examined in the course of his research omitted at least one of the essential elements which were to be covered under those regulations.

7.19   Mr Waite found that while safety cases described the arrangements for the management of safety, few of them demonstrated how those arrangements were suitable and sufficient for the purpose. He referred to this as

"...constructing an argument to give confidence that they have considered all the risks".

He acknowledged that this had been addressed to some extent in the changes which had been made by the 2000 Regulations, such as the introduction in para 16 of the First Schedule of the requirement to include in the safety case the duty holder's proposals for improvements to the safety case and the health and safety measures referred to in it. He and Mr Brown made a number of other criticisms of safety cases, including that they tended to be compliance-driven to satisfy the acceptor, rather than working documents to improve safety controls. It is important, as Professor Evans pointed out, that the regime should encourage people to think as actively as they can to reduce risks.

7.20   One matter which is not explicitly stated in the regulations is that a safety case should demonstrate that the risks with which it is concerned have been reduced ALARP in accordance with the statutory duty of the duty holder. This is, however, suggested in the guidance issued by the HSE. Dr R A Cox, Consultant Engineer, said that the principal shortcoming of the regulations was that they did not require such a demonstration or of compliance with any other risk targets. He said that, instead of being used as a defence for inaction, ALARP should have been given a rigorous interpretation in every instance: "It is a philosophy that leads to continuous improvement". However, Professor Evans found that none of the 22 railway operators whose safety cases he had examined was able to demonstrate that the risk which it imported satisfied the ALARP test. In my view a safety case should show that the duty holder has reduced the risks associated with its operation ALARP. Achieving and maintaining such a reduction is required by health and safety legislation. However, I do not consider that it is necessary for the detail of the examination, assessment and control of individual risk to be set out in the safety case. There is an existing tendency for safety cases to become bureaucratic and I have no wish to encourage that tendency. It should be sufficient if the safety case points to the methods which have been used and to where the details can be found.

7.21   Mr Waite saw it as a positive feature that the Safety Case Regulations were concerned with all risks, as a large proportion of injuries were caused in relatively minor accidents. Since all risks were covered, effort could be put into different areas in proportion to the degree of risk which was involved. Mr K Bird, Chairman of c2c Rail, described in evidence how the advent of safety cases had made TOCs understand in a structured way the risks which were created by their operations. Risks could then be categorised by reference to their importance and the likely exposure to them. As regards the assessment of risks, Mr Waite pointed out that there was much less use of predictive techniques than in other industries, so that the focus was inevitably on historical data. Major catastrophes were fortunately infrequent. He said that the use of predictive techniques which had been developed in other major hazard industries was useful in anticipating them and designing preventive measures.

7.22   The Safety Case Regulations do not require the duty holder to comply with Railway Group Standards. Nor do they require it to set out its arrangements for complying with them. Mr Muttram said that Railtrack had not accepted a safety case in which there was not such a commitment. The basis for this assertion was not clearly made out in the evidence. In particular it was not clear whether this was based on a direct commitment by the duty holder or whether it was based, in a more roundabout way, on the duty holder stating a commitment to act in a manner which was consistent with Railtrack's safety case, in which an undertaking to comply with Group Standards could be found. I recommend that duty holders should be under a statutory duty to comply with Group Standards in so far as they relate to matters of health and safety.

The preparation of safety cases

7.23   As may already be clear, the discipline of producing a safety case has an important value in itself. This was summed up by Mr Brown who said:

> "It is actually the process of writing the safety case, the determination of the risk, the consideration of the management control systems to be put in place and the

ways in which those risks will be controlled within that management system that are at the heart of what is most important about a safety case".

The use of outside consultants to produce safety cases was, in his words,

> "...completely ineffective. I think if people do not actually do this process in-house and do not involve all parties in it, it will not work. And I have got personal experience of that".

Mr Tunnicliffe informed the Inquiry that the process of developing LUL's safety case was "management intensive and uncovered many weaknesses". Likewise Mr Waite saw inherent value in the learning opportunities for management and senior technical people in the organisation, particularly when carrying out risk assessment and setting out the safety management systems. He too stressed the importance of safety cases being written by those within the organisation. The Entec report stated:

> "If employees understand why they are doing what they do they are more likely to apply the principles and less likely to violate rules".

As was pointed out by Mr W A T Alder, Principal Consultant for Entec UK, a learning process is consistent with the goal-setting approach which is embodied in the safety case regime.

7.24   ATOC agreed that the process of producing safety cases entailed the close involvement of line management with safety professionals. Mr Baker, who described the way in which his company's safety case had been prepared, described the safety case itself as

> "...a practical, living document which helps improve safety, in the same way as safety audit is not a painful process but is a very useful learning experience".

His concern was that there were some organisations where the safety case was of too high a level to be of practical value.

7.25   In this context I noted that the Rail Unions drew attention to failures to consult safety representatives, as required by Regulation 9(11) of the 1994 Regulations, which is now replaced by Regulation 14(8) of the 2000 Regulations. Mr R Rosser, General Secretary of the TSSA, spoke of

> "...the involvement of people who are there on the ground and know actually what is going on, will know about, for example, if there are difficulties in particular areas that need to be addressed, know the reality of the situation as opposed to the theory of what it should be like".

Once again it is useful to refer to the Entec report, which stated:

> "If employees are involved in producing the safety case (rather than just being told about it) they would have 'ownership'. This can bring stronger commitment to the arguments".

Mr Brown remarked that failure on the part of management to ensure that "the message gets through" to all levels was

> "…very much related to the failure to involve everybody in the process and very much the failure of constructing documents that people could find accessible and understandable and, crucially, helpful".

Likewise Mr Tunnicliffe said that LUL found that the key process, and one of the key features of the safety case, was the involvement of employee representatives. They had been surprised at just how valuable was their input.

The uses of safety cases

7.26   One of the descriptions of a good safety case which has become familiar is that it is a "living document". Thus Mr Brown said:

> "Whatever the level of detail, by placing the safety case in a live management system so that it becomes part of that system makes it more likely to be maintained as a current document and assist decision-making by the duty holder".

In the HSE's review "The Management of Safety in Railtrack" the safety cases of TOCs were described as living documents

> "…in that they reflect the daily operation and management of risks by the TOC and as such should be amended to reflect any changes and subject to regular review to ensure that they remain valid".

Mr Waite said:

> "This is to keep the actual practice of the company in step with the description so that there is no ambiguity about what is accepted practice by the company management".

However, in their report Entec stated that, while risk assessments were relied on in safety cases, they were not updated to take account of the occurrence of accidents and the lessons learned from them.

7.27   In the words of the HSE's draft guidance on the 2000 Regulations, the safety case serves the purpose of providing

> "…a comprehensive core document, with links to other more specific documents, rules and procedures, against which management and HSE can check that the accepted risk control measures and the health and safety management systems have been properly put into place and continue to operate in the way in which they are intended".

Professor Evans considered that safety cases provided some sort of benchmark against which the company could be judged, although he had some reservations as to whether they were sufficiently defined for this purpose as opposed to providing a more general indication of the matters which ought to be assessed. Mr Muttram observed that there was some variation in the extent to which safety cases were treated as working documents. Mr Baker expressed doubt about whether the Railtrack safety case was usable in that way. Professor Evans also indicated a number of ways in which a safety case could deal with change. It helped to provide a level of continuity which would survive the loss of safety personnel, and it also helped to control changes. I referred earlier to para 16 of the First Schedule to the 2000 Regulations which requires duty holders to identify areas in which they see that their systems are deficient and to set out timetables for changes. Mr Brown said that where this was done it would effectively become an agreement between operators and the HSE, leading to a process of continual improvement.

7.28   Mr Brown also gave evidence that a safety case could be used as an aid to training. Companies could use it to highlight key areas which required training, and then as a basis for reviewing whether personnel were currently performing to that standard. The descriptions of company structures as set out in the safety case could also be used to train employees at all levels of an organisation. Mr Waite indicated that it would be helpful if a non-technical summary of a safety case could be prepared in order to explain the argument for safety without too much technical detail. This would also help employees to understand the company's safety case, as it would highlight key points in simple and comprehensible language. I note that the HSE recommend in their "Assessment Criteria for Railway Safety Cases", which was published on 21 November 2000, that a safety case should contain an executive summary which briefly outlines how and where the main safety arguments are set out in the safety case, any changes that have been made resulting from reviews and revisions to the previous version.

7.29   Mr Brown made three suggestions for increasing the understanding and knowledge of employees of their company's safety case. These were:

(i)   the availability at depots of electronic access to the text of the safety case and relevant company documentation;

(ii)   the use of the briefing process to explain key parts of the safety case to drivers, signallers and other employees; and

(iii)   the use of clear, easily understandable summaries explaining key parts of the safety case.

I understand from the account of the Inquiry's seminar on Employee Perspectives on Rail Safety that GNER have produced a synopsis of their safety case for their employees. Mr Baker said that Northern Spirit had produced an A5 version of their safety case which was supplied to middle and junior management. It appears that similar actions are being taken by other TOCs. ATOC emphasised that staff need to understand not only the rules but also how these rules contribute to their environment; and to understand why their individual duties are important and how they relate to the whole.

The assessment of safety cases

7.30   Mr Waite suggested that part of the process of safety case acceptance should be checking that a system as described in the safety case was actually in place. On this basis the audit would concentrate more on how that system was working in practice and how it was ensuring and improving safety. I support this approach.

7.31   Mr Coleman informed the Inquiry that in future the HMRI would be looking for

> "...very much improved safety cases...better definition of management issues, the whole safety management systems, responsibility of individuals, definition and control of interfaces".

He also pointed out that Regulation 8 of the 2000 Regulations introduced a power on the part of the HSE to require a duty holder to revise its safety case. This could be used if the duty holder was not persuaded by other means to improve it.

7.32   Mr Brown is responsible for a new team of inspectors which will seek to emphasise the need for management commitment, training and motivation. His command of the subject and his commitment to the work were plain. He indicated the approach which he would prefer to see safety cases taking in the future. In one model the duty holder described what was done, along with the development process, indicating its vision of what it was aspiring to achieve. In the other model – which he preferred – the duty holder described what should be done, but pointed out the present deficiencies in achieving that aspiration, with an explanation of the procedures to deal with them. He said that the acceptance process should be one of continuous assessment. The criteria used should be as unambiguous as possible and freely available to duty holders.

7.33   As I have already noted, the HSE have published a detailed set of criteria for the assessment of RSCs. Mr Brown explained that they had been placed in the public domain not only to allow the HMRI to refine them in the light of the views of duty holders, but also to enable duty holders to know what would be done in the assessment of their safety cases. It was certainly a regime which would make greater demands on the HSE. It would also require more of the duty holders. However, ATOC regarded the procedures for safety cases which are set out in the 2000 Regulations as highly bureaucratic. ATOC expressed concern that the HSE's guidance demonstrated that an "extreme level of prolixity" would be required.

Auditing

7.34   Auditing is a vital component in both the management and the regulation of safety. Mr Tunnicliffe observed:

> "In practice human beings are imperfect and deviations from the expected will occur. It is therefore essential that a comprehensive system of audit stands alongside the safety procedures to assure responsible management that they are indeed taking place. Once again, in practice, they will not in some cases, and it is essential that this is grasped as a learning experience, to refine and correct".

Audit is on the one hand a quality assurance exercise, and on the other a compliance process. However, as was pointed out by Professor Evans, the amount of detail which is provided by the safety case may determine the extent to which it can be used as a basis for auditing.

7.35   A number of witnesses, including Major Holden, drew attention to weakness in auditing. In his written statement he said:

> "My concern has been that there has been a lack of penetration in the audits, which have tended to chase paper trails rather than check that what should be going on on the ground is, in fact, going on. This lack of penetration may, in part, be due to the lack of skill of the auditors but it may also lie in the belief that all that is required is a pure compliance audit of the accepted safety case. The vital question as to whether or not the safety case itself is adequate and appropriate to the circumstances is seldom asked".

Internal auditing

7.36   Under paragraph 5(d) of the First Schedule to the 2000 Regulations the duty holder is required to set out particulars of the arrangements that it has established for the carrying out of audits, for the making of audit reports and for the review of health and safety performance.

7.37   The remarks which I have quoted in para 7.35 are particularly apposite in the case of internal auditing. As Dr I A P Scott, Director of Safety, Health and Environment for Eurotunnel, pointed out, internal auditing carries with it the benefit that internal auditors are able to keep regularly in touch, and see what is happening and where to look to make improvements. Mr I M Waldram, Past-President of the Institution of Occupational Safety and Health (IOSH) and an experienced health and safety advisor, emphasised that it was vital that each company itself operates a robust internal audit system, as opposed to relying on an external audit body such as Railway Safety. He added that he found that, in general, well-trained internal teams could identify deficiencies which were missed by external auditors:

> "This is because the internal teams know the people and general organisation so much better, so are less easily fooled by processes which look good on paper but aren't easy to implement in practice".

I find the observations of these witnesses to be compelling. They mean that if auditing is to do its job properly, it should be both "top down" and "bottom up". This is clearly to be recommended.

7.38   The Inquiry heard evidence that certain weaknesses were found in the internal auditing of TOCs, ROSCOs and IMCs by DuPont Safety Resources and Entec. However, the position appears to be improving. Mr Baker described how changes had been made. These were in line with the principles set out in the HSE guidance "Successful Health and Safety Management", which emphasised the importance of organisations having systems to manage health and safety on a day to day basis. It may be noted that he said that ATOC had not issued guidance to their members about internal audits. This was because it had not been the role of ATOC to become so directly involved in safety management. That could lead to further confusion. There were already too many bodies.

<u>External auditing and enforcement</u>

7.39   Under Regulation 9(1)(b) of the 2000 Regulations Railtrack are required to procure regular audits from an assessment body, in practice Railway Safety, of train and station operators "in relation to railway infrastructure" which is in their control. This is in addition to the external audits, monitoring and checking which are carried out by or on behalf of Railtrack, in connection with the discharge of their duties under the Health and Safety at Work Act, of those who operate, or supply products and services for use, on the infrastructure. In their submissions Railtrack drew attention to the fact that in the HSE's report "The Management of Safety in Railtrack" the S&SD were complimented on their processes for the acceptance of safety cases. However, the report went on to observe that

> "…the diligent assessment of TOC RSCs is not supported by adequate arrangements for ongoing monitoring of performance of TOCs against their safety case commitments".

The report listed a number of areas of concern where there should be improvements, including the fact that deficiencies identified by previous audits remained unresolved. Asked whether this was still a fair observation at the time when he gave evidence on 27 November 2000, Mr Muttram said that he believed that there had been a significant improvement, but that he could not give an assurance that every audit action was followed through.

7.40   Mr Muttram said there had been a general improvement in the audit procedures of the S&SD. He said:

> "We would look for evidence that it (the operator) is working to that safety management system that it is committed to, and that that safety management system is working in practice".

He explained that this was why in 1999 the S&SD had introduced a process of auditing both "top down" and "bottom up". A second auditor would work at ground level within the company. He would, for example, accompany driver standards managers in the driving cab, talk to drivers and go into the maintenance depot and look for evidence. It may be noted that this approach was not adopted in the auditing of Railtrack Line, on the ground that most of their work was done through contractors. Mr Muttram said that the S&SD had developed a Railway Safety Case Compliance Audit Protocol. However, it was pointed out that the latest issue of the protocol did not appear to reflect a "bottom up" approach. Mr Coleman said that he would like to see auditing looking not only at the processes that were in place, but also at whether they were delivering safety, but he accepted that he had not made any unfavourable comment on the protocol in this respect.

7.41   "The Management of Safety in Railtrack" also drew attention in February 2000 to Railtrack's lack of a clear system, supported by clear criteria, for successively stronger degrees of action in cases of breach. It stated that this

"...results in Railtrack exercising only limited control of the TOC once the RSC had been accepted".

Mr Muttram said that there had been an improvement in escalation procedures. He said that for minor breaches, particularly those identified through auditing and monitoring, the S&SD required the operator to produce an action plan and identify corrective actions to be carried out within specified timescales. For more significant breaches the S&SD could, in addition, demand that transitional controls be put in place. Where the breach was serious and created a serious risk to safety, Railtrack advised the HMRI. However, Railtrack contended there was still a need for a clear and more formalised hierarchy of powers which would be exercised by Railtrack and Railway Safety. On the other hand, Mr Coleman said that the HMRI had never accepted the claims of Railtrack Line that they had no power apart from "the nuclear option" of stopping an operator from using the network. He referred in particular to the regulation in regard to co-operation, which is now Regulation 11 of the 2000 Regulations. To these have been added the new Regulation 12 which imposes a duty on Railtrack to take steps to ensure that an operator complies with its safety case where that concerns Railtrack's duties; and Regulation 13 which requires the HSE to be notified in certain cases of non-compliance.

7.42 It also has to be noted that the auditing of holders of safety cases by Railway Safety is limited to ensuring that the introduction of risk to the infrastructure is effectively managed and reduced. It does not cover such matters as the safety of the interior of rail vehicles or maintenance depots which are within the exclusive occupation of a particular TOC. Mr Baker assured the Inquiry that his company did not fail to cover such matters in its internal auditing, and he expected that other train operators would control those areas in much the same way.

7.43 In their report on the Management of Safety by Railtrack, DuPont described examples of excessive auditing, or "death by auditing". Mr Muttram said that he had some sympathy with the views of DuPont. It was one of the reasons why he considered that contractors should be brought within the safety case regime. There had been an effort to improve co-ordination between auditing by the S&SD and auditing by Railtrack Line, with a view to avoiding unnecessary duplication. The evidence demonstrated to me that one of the ways in which an external audit can be used to best effect is not to duplicate the work of internal auditing but to scrutinise the outcome of that work. Mr Muttram described the S&SD approach as an office-based exercise which looked for evidence that the internal audits had taken place and that what had been shown to be deficient in those audits was in fact followed up. However, it is not clear whether the S&SD reviewed the effectiveness of the internal audit procedures themselves. This, in the submissions of the Collins Passengers' Group, underlined the importance of the operator ensuring that its internal auditing was effective as "the primary audit resource for railway safety cases".

7.44 The external auditing of railway operators is carried out not only by Railway Safety, in succession to the S&SD, but also by the HMRI. The function of the latter is to test the compliance of companies not only with the requirements of the safety case legislation but also with health and safety legislation at large. The foundation for the activities of the HMRI lies in Section 18 of the 1974 Act which places the duty of enforcing health and safety legislation on the HSE. The processes of external auditing to which I have referred involve the duplication between the function of Railway Safety and that of the HMRI. It is not a complete duplication, of course, because the function of Railway Safety is limited to a concern with the importation of risks to the infrastructure, whereas the function of the HMRI is with any risks within the operation of the railways. A General Issues Report by the HSE, which dealt with a number of matters arising from the internal inquiry in the aftermath of the crash at Ladbroke Grove, acknowledged that there had been weaknesses in the performance of the HMRI in regard to safety cases. Some inspectors had seen auditing against a safety case as a complex affair: it was work with which they had not felt comfortable. Field inspections had not routinely been used to test the validity of the safety case, and there had been little co-ordinated follow-up of issues arising. Since then significant improvements have been made in the internal processes of the HMRI. The report also recorded that enforcement by the HSE of compliance with safety cases had been ineffective. This appears to have been due to lack of sufficient resources (the requirement for resources having been on the basis of a "light touch" approach), and insufficient account being taken of experience in other safety case regimes.

Public assurance

7.45   The draft guidance to the 2000 Regulations which was issued by the HSE stated that other main purpose of a safety case is

> "...to give confidence that the operator has the ability, commitment and resources to properly assess and effectively control risks to the health and safety of staff, contractors, passengers and the public".

There was conflicting evidence as to whether there was sufficient interest or understanding on the part of members of the public to make it worthwhile for copies of RSCs to be available for public inspection as is now required by Regulation 14(1)(c) of the 2000 Regulations. However, there appeared to be substantial support for the view that the fact of their availability for public scrutiny would be of benefit in generating public confidence. Counsel for the bereaved and injured represented by the Southall and Ladbroke Grove Solicitors' Group suggested that safety cases should be posted on the websites of rail operators or their trade association.

7.46   Regulation 14 of the 2000 Regulations does not appear to require that copies of the audits of railway operators should also be made available to the public. Major Holden considered that there would be merit in making them available, and he could see no reason why they should not be available to persons who had a proper interest. Sir David Davies, President of the Royal Academy of Engineering, said that it was envisaged that the reports of audits by Railway Safety would be published. Mr R J Morris, Executive Director of London South East and formerly the Technical Director of Safety and Operations for the SRA, indicated that making an operator's processes more public and transparent was likely to improve them, and he supported making audits public. Mr Brown was sure that they should be.

An extended safety case regime, accreditation and licensing?

7.47   A number of parties and witnesses advocated the extension of the safety case regime to include companies to which the requirement to hold a safety case does not at present apply or does so only to a limited extent. The discussion came to form part of a wider discussion of the "accreditation" of companies. Since such a term may be of uncertain meaning, I should explain what I have in mind. A company is accredited if an external body has determined that it has suitable processes for controlling the safety of the products or services which it provides, in particular where they are safety-critical. In paras 7.72-7.73 I set out my conclusions in regard to this topic, on which there was, unfortunately, no agreement between the parties.

7.48   It may be helpful if at this stage I outline the current means which are adopted for the qualification of the suppliers of safety-critical engineering products and services in the rail industry under Railway Group Standard GM/RT240, which was issued in December 1995. That Group Standard states that the purpose of requiring qualification is to minimise the importation of safety risks. Examples are the procurement of track, structures, vehicles, rail vehicles, components, systems, telecommunications, signalling, repairs, modifications, maintenance services, engineering activities, hire of labour, consultancy activities and technical advice. It may be noted that the Group Standard:

   (i)   does not involve a determination by an external body. The person or organisation which procures the supplies has the responsibility of qualifying the supplier on behalf of his or its employer;

   (ii)   does not prescribe in any detail what is adequate for the purpose of qualification. There requires to be a process for qualifying suppliers which establishes certain minimum requirements, including that the supplier understands the risks to the railways which are associated with the product or service, is competent to produce a safe product or service, and has a quality and safety management system which is adequate and follows the principles of national and international standards, and is competent to evaluate and select sub-contractors. The procurer or his agent is to have auditable documented processes which demonstrate continual compliance with Group Standards; and a process in place for surveillance of the integrity of a supplier's existing qualification, particularly when new contracts are let; and

   (iii) is not concerned with risks other than those imported to the infrastructure.

7.49   The parties referred to a number of different types of company which provided products or services to the rail industry. The principal examples are discussed in what follows.

7.50   ATOC proposed a system of accreditation for all service providers and that safety cases should be produced by some of them. The requirement for a safety case should be determined by the nature of the activity of the particular organisation. If it was, or could affect activities which were, safety-critical, it needed the internal discipline and external scrutiny of a safety case. Mr Eccles explained, by way of example, that those who were responsible for the maintenance of a signalling system, electrification system or track should be accredited through a safety case. The system which was advocated by ATOC would be administered by the National Rail Safety Authority (NRSA), a new safety regulator which they proposed. ATOC's case was that the present requirement that operators had to be satisfied with the safety and integrity of services and products used by them involved duplication.

7.51   Railtrack submitted that it followed from a holistic approach to the assessment and management of risk that all parties whose activities were integral to the maintenance and operation of the railways should be subject to the safety case regime. The activities of IMCs were fundamental to the safe operation of the railways. At the same time Railtrack said that they acknowledged their responsibility for proper supervision of contractors. This was shown by their requirement that contractors should produce assurance cases. There would be little practical difficulty for IMCs if the statutory regime were extended to them. Railtrack submitted that their proposal carried a number of benefits. First, it would end the anomaly by which an IMC was treated as a train operator only to the extent, if any, that it operated steel-wheeled rail vehicles whether in or out of possession. All IMCs which operated such vehicles outside possessions should be required to produce a statutory safety case covering all their activities. This would cover all the major contractors. Secondly, this would strengthen and simplify the audit processes, and in particular lessen the number of audits. Thirdly, it would provide greater clarification of the responsibility of IMCs; and, fourthly, it would help to develop a good safety culture. Mr Muttram said that he regarded the absence of a requirement for a statutory safety case in the case of contractors as a deficiency of the present regime. He also said that contractors were not subject to a duty to co-operate under the regulations. However, this latter statement does not appear to be accurate in view of the terms of Regulation 11(2)(b) of the 2000 Regulations, previously Regulation 8(2)(b) of the 1994 Regulations, which imposes a duty to co-operate on

>   "…an employer of persons or a self-employed person carrying out work on or in relation to premises or plant owned or controlled by the duty holder".

7.52   Counsel for the bereaved and injured represented by the Southall and Ladbroke Grove Solicitors' Group submitted that statutory safety cases should be required of both IMCs and TRCs, and accordingly subject to a compulsory annual statutory audit. Other suppliers should be subject to a system of certification, under which they would be inspected, and could be checked when their certification was being renewed. The standard to be applied, and the degree of scrutiny, should correspond with the degree of risk involved in the activities of the supplier. Counsel submitted the proposition that

"...legal responsibility should correspond more closely with factual control than under the present nominalist structure".

Counsel for the Collins Passengers' Group took a similar line.

7.53 Amey Rail, who are an IMC, advanced two objections to the proposal that IMCs should be required to provide a safety case. The first was that it would lead to a situation in which more than one member of the Railway Group would have direct responsibility for the safe maintenance of the infrastructure, whereas only one of them would be in effective control of that maintenance. Railtrack Line controlled the availability of possessions and the process of renewals. It followed that an IMC could be in breach of its safety case for reasons outwith its control. Amey Rail sought support from the evidence of Mr Coleman. He was not in favour of an IMC requiring to hold a statutory safety case as this might confuse Railtrack's primary responsibility for the management of their own business. Railtrack had the freedom to choose their own contractual arrangements. It was inappropriate that the safety regulator should become involved in the question of the adequacy of these contractual arrangements. Amey Rail also founded on the evidence of Mr Muttram when he said:

"Railtrack believes that any further intrusion which fetters, or seeks to direct, the terms on which Railtrack employs contractors may lead to split responsibility and accountability".

A practical example was provided by Mr Carr when he said in evidence:

"We could argue that there needs to be a different system of possessions to enable us to do our work and that could be in conflict with Railtrack's. Now...it is Railtrack's responsibility to ensure that our concerns and our requirements dovetail into the requirements of the train operating company".

The second objection was that there could be confusion as to the interpretation of Group Standards. Mr Carr pointed out that under the current system, new or revised Group Standards were the subject of variation instructions under the maintenance contracts. These allowed Railtrack to interpret the Group Standards by issuing a Line Standard. If IMCs were to become the holders of statutory safety cases they would "automatically" be required to implement all such Group Standards. This could lead to a possible conflict of interpretation. Amey Rail submitted that any difference could not be resolved by the standard-setting body. They submitted that there were sufficient control mechanisms at present to ensure that the operations of IMCs were conducted safely. They drew particular attention to the development of a more collaborative approach, which should facilitate Railtrack's ability to monitor and audit their performance in an effective way.

7.54 Mr C J Wheeler, Project Manager of the National Track Safety Strategy Group and Chairman of the Association of On-Track Labour Suppliers, said that it was absurd that a contractor who undertook the clearance of forestry scrub had to hold a safety case if it operated a JCB which had steel wheels. The time spent in complying with, in his words, "all the full panoply of railway safety case requirements" could be better spent in sorting out safety matters involved in the operation. On the other hand he suggested that safety-critical work should be performed only by contractors who had been approved in regard to their quality and competence for such work. At present such work could be sub-contracted by contractors who did not hold safety cases and did not have detailed knowledge of railways.

7.55   The Joint Rail Unions opposed the proposal made by ATOC and Railtrack for the extension of the requirement to hold a safety case. Their opposition was based on a number of grounds. The first was that the upkeep of the infrastructure was an integral part of Railtrack's responsibilities and fell entirely within their control. Railtrack chose to continue to put out such work to contract and they chose who should do such work. Accordingly they should continue to be fully responsible for any importation or risk that this might entail. Secondly, the proposal introduced a legal concept that infrastructure control and infrastructure maintenance support/renewal were segregated activities. Thirdly, it would not provide any added value. There was no difference between the difficulty in terminating a contract on the ground of non-compliance with a contractor's assurance case and the difficulty in terminating it on the ground of non-compliance with a statutory safety case. Fourthly, it was not correct that a main contractor was "outside the statutory fold". A significant breach by it of its safety assurance case would invariably constitute a breach of the 1974 Act or other relevant statutory provisions.

7.56   In response Railtrack maintained that the opposition to the proposal ignored the fact that the employer of contractors would still remain under a legal duty. It also ignored the possibility that the employer's duties could be strengthened by an amendment of its own safety case. Further, Railtrack maintained that there was no reason why contractors should become more likely to interpret Group Standards differently from their employer. It could not be said that the imposition of a duty to produce a statutory safety case was excessive since contractors already required to produce assurance cases. There were valuable benefits in bringing contractors within the same regime as their employers. This had the benefit of enhancing the safety culture and improving the focus on the control of safety.

ROSCOs

7.57   Mr Abbott explained that since 1997-98 his company, Angel Trains, had been accrediting and auditing suppliers, and some sub-suppliers, for all products supplied to it, and not merely those which were covered by Group Standard GM/RT2450. He understood that the practice of other ROSCOs was broadly similar. He was uncertain as to whether his company required to be qualified under that standard, but it supplied its customers with details of its processes, including its safety management systems, along with its accreditation programme as it was clearly supplying safety-critical services. However, his company had never been audited by the TOC under the Group Standard.

7.58   The position of the ROSCOs was that they accepted the need for them to be accredited, but did not believe that they should be required to produce a safety case as this would blur the distinction between those engaged in operational activity and those who were in "the second tier". In their closing submissions they stated that the idea of a "blanket" safety case sat uneasily with the operational slant of the present safety case regime. A tiered approach made more sense. They pointed out that once a ROSCO had delivered a rail vehicle to an operator, either after its construction or following heavy maintenance, it had no control over the day to day operation of that vehicle and hence could not control the risk arising from its use. Accreditation should not be allowed to dilute the fundamental principle that those who imported risk on to the infrastructure should be responsible for controlling it. However, suppliers should demonstrate their ability to manage the risks within their own sphere of operation. It was not expected that in practice there would be any difference between the demonstration of safety management systems required for the purposes of a safety case and that required for the purpose of accreditation. This was in line with evidence which had been given by Mr Abbott. On the basis of these submissions the ROSCOs maintained that the case for their accreditation should be restricted to the activity of heavy maintenance and should not go beyond the point where a train was returned to a TOC for running on the network.

7.59 ATOC submitted that the distinction which the ROSCOs had drawn between operational activity and the activity of those in "the second tier" was misconceived both legally and logically. To treat risk as generated primarily by train movements rather than by total system risk made up of the activities of different companies and the interactions between them was a mistaken philosophy. The need for safety management systems was dictated by the level of risk presented by the operations of each.

7.60 Railtrack took the same approach to ROSCOs as it did to IMCs. Those who procured, controlled, or carried out heavy maintenance on, rolling stock should be required to hold a statutory safety case. This should result in better and earlier identification of hazards and assessment of risk, especially at interfaces. Those who were subject to the safety case regime would obtain the benefit of greater involvement in the setting of Group Standards and the ability to exercise the right of appeal.

7.61 The Inquiry was also asked to consider whether not only heavy maintenance but also the responsibility for the rectification of design faults and one-off major faults throughout the lifetime of trains should attract the need for a safety case. It was suggested that the same could also apply to their upgrading and modification.

Manufacturers and suppliers

7.62 Professor R J Kemp, who is the Safety Director of Alstom Transport and who gave evidence for the interests of the RIA, said that he was in favour of manufacturers being accredited rather than having to produce a safety case. It would be non-statutory. He observed that if a TOC was responsible for a train which it wished to run on the network, it had the responsibility of demonstrating its safety. It would simplify the position for the TOC if it could say that it had satisfied itself by using a certified train from an accredited manufacturer. At present manufacturers required as a matter of contract to produce technical "product safety cases" to justify to Railtrack that a product or system would not import risk. He was not in favour of manufacturers requiring to hold safety cases, because they were not carrying on the same unitary activities as the present holders of safety cases, and because their products were distributed to different countries. On the other hand, he accepted that the holders of safety cases had the benefit of having direct access to the process for the approval of Group Standards and could appeal against a proposed standard. But in support of the position which he took, he pointed out that modifications might be made to a product without reference to the manufacturer. He accepted that the RIA took the view that manufacturers should have a continuing responsibility for the product. If that was so it was desirable that manufacture should be subject to greater regulation. However, that also placed responsibility on the operator not to make fundamental changes to the product without consulting the manufacturer: "The additional responsibilities cannot just be one-sided".

7.63   Mr Eccles, in setting out the position of ATOC, explained that their view was that accreditation would cover the suppliers of vehicles, parts and equipment. Anyone who wished to do business in the rail industry would have to pass through that process of vetting and approval.

7.64   Counsel for the bereaved and injured represented by the Southall and Ladbroke Grove Solicitors' Group submitted that since manufacturers were increasingly in control of the safety of products throughout their life, some form of safety case or certification should apply to them. All the risks in relation to the system, and not merely the risk imported to the infrastructure, should be covered.

Licensing

7.65   Before referring to the discussion about the licensing of individuals it is convenient to refer to the existing arrangements. Regulation 3(1) of The Railways (Safety Critical Work) Regulations 1994 provides that no employer is to permit any of his employees to undertake any safety-critical work unless:

"(a)   the employee is competent and fit to undertake that work;

(b)   there is in existence an adequate record of any relevant assessment undergone by that employee;

(c)   the employer has issued the employee with a means of identification; and

(d)   the employer has established suitable arrangements to enable a railway operator who is affected by the work to examine the record referred to in sub-paragraph (b) above, or to be informed of its contents".

The expression "safety-critical work" means, in terms of Regulation 2(1), work by a person:

"(a)   as a driver, guard, conductor or signalman or in any other capacity in which he can control or affect the movement of a vehicle; or

(b)   in a maintenance capacity or as a supervisor of, or look out for, persons working in a maintenance capacity...".

It may be noted that these regulations:

(i)   place the onus on the employer to ensure compliance. No external body is involved; and

(ii)   no standard is laid down against which fitness and competence should be assessed.

7.66   This set of regulations is supported by an HSC Approved Code of Practice and general guidance issued by the HSE. Railway Group Standard GO/RT3260, the second issue of which was in August 1998, defines the scope of the work which is to be considered safety-critical and the requirements for employers' systems for ensuring competence and fitness of persons required to carry out such work. This is supported by a non-mandatory Railtrack Code of Practice. The Inquiry was informed that the National Competency Control Agency had been set up by Railtrack to record those individuals who had been suitably trained in personal track safety and other qualifications for safety-critical work. About 100 training organisations had been accredited for this purpose. The Agency issued cards called "Sentinel" cards to the qualified individuals. There were about 100,000 such individuals, sponsored by about 2,000 companies.

7.67   Most of the discussion was concerned with the possible licensing of drivers. TOCs are responsible for creating systems, processes and procedures to comply with a Group Standard on train driving (GO/RT3251) and the Group Standard to which I have referred in the previous paragraph using the guidance contained in Railtrack Codes of Practice. Under the Group Standard on train driving TOCs are required, at a frequency determined by each individual TOC, to review the effectiveness of the system in place to deliver the level of competence required. Best practice is set out in an ATOC Code of Practice.

7.68   Recommendation 7 of Professor Uff in his report on the Southall rail crash was that Railtrack and ATOC should establish a national "qualification and accreditation system for drivers". In this Inquiry Mr Muttram said that he was in favour of driver licensing and would like to introduce such a scheme. Railtrack had supported the conducting of a full review of the competency certification of all safety-critical workers. HSE had the lead in that action in reviewing the Safety Critical Work Regulations. Sir Alastair Morton, Chairman of the SRA, also indicated that the SRA had supported the idea that drivers should have independent accreditation. In my report on Part 1 of this Inquiry I recommended that a current study by ATOC on the central licensing of drivers should be progressed expeditiously (Recommendation 15). During the course of Part 2 ATOC produced a position paper on that subject. In the paper they drew attention to the fact that there was already a de facto system of licensing which was undertaken by each TOC. (In the paper the same remarks applied also to FOCs.) A driver could not drive unless he had been passed as competent in his understanding of the Rule Book and local operational knowledge. Much of the latter, which was concerned with traction and route knowledge, was invariably specific to the operation and service pattern of the individual TOC. There was some scope for central licensing in regard to driver knowledge of the Rule Book, the testing of which could be undertaken and recorded centrally. Alternatively, testing could be delegated to an accredited local provider or TOC. This would form a first element of a database creating a Competent Train Crew Register. A national database of drivers competent in local operational knowledge was considered to be of little practical value, given the specificity of such knowledge. The benefits of such a system would be the application of a national standard and the creation and maintenance of the register. It was also consistent with the system of licensing administered by the CAA, which was favoured by ATOC as a model for the rail industry. The disadvantages were that national testing might discourage companies from requiring a higher standard. It would involve additional cost and the redundancy of current testing facilities.

7.69   In their submissions ATOC submitted that while there might be some merit in central licensing, it could not replace the essential training in traction and route knowledge provided by the individual employer.

7.70   The Joint Rail Unions supported a national licensing system for drivers, linked to the obtaining of NVQs. They regarded this as important for the restoration of public confidence. High national standards embodying "best practice" should be mandatory. Meeting those standards should be essential to qualify an individual to drive a train. There were some advantages in centralised training, such as economies of scale, the accumulation of expertise and uniformity of practice. They saw an important role for the RITC in setting national standards and supervising assessment procedures. They stated, without any apparent reservation, that

> "...the national standard for any national train driver's licence must incorporate robust retention of route and traction knowledge".

Test results would form the first element of a database creating a Competent Train Crew Register. Contrary to the view of ATOC, a national database of drivers competent in local operational knowledge – traction, route and sectional – could be established.

7.71   Counsel for the bereaved and injured represented by the Southall and Ladbroke Grove Solicitors' Group submitted that the setting up of a national accreditation system for drivers, as recommended by Professor Uff, should be extended to others who were engaged in safety-critical work, and beyond that to others whose work involved "a significant degree of risk". This was necessary for public assurance. It would be a function of the NRSA, which they advocated, to determine the classes of work to which the system would apply, along with the courses, tests and re-tests which would be required. They emphasised that various factors might lead to particular employers not acting in accordance with good practice. They pointed out that railways were the only form of transport where there was no strong independent agency involved in licensing. In the light of the evidence it was also submitted that the Sentinel card scheme, which was accredited only by Railtrack, had drawbacks and was subject to possible abuse.

Conclusions

7.72   The submissions that the safety case regime should be extended plainly involve a change in the basis for the Safety Case Regulations. To date the regulations have been concerned with the control of risks for which "railway operators", namely infrastructure controllers and the operators of trains and stations, are responsible. The primary justification for the regulations has been the complexity of the interfaces between these railway operators. The proposals which were put to this Inquiry took as their basis the control of risks for which the suppliers of products or services for the railways are responsible, as part of the total risks which may affect railway safety. There is some attraction in the idea of extending the requirement to hold an accepted case to those whose activities take place on and affect the safety of the railways (i.e.

contractors) and those who have an ongoing responsibility for the safety of what is used on the railways (e.g. ROSCOs, at least in regard to their responsibility for heavy maintenance). However, on the evidence before me, a number of considerations seem to me to detract from the advisability of such a proposal. To require a safety case not only from the railway operators but also from the parties who are in contractual relationship with them seems to me to involve the risk of duplication and confusion. There is a danger that it may be treated by a duty holder as relaxing its responsibility for controlling risk on the railways. The proposal seems to me to be somewhat bureaucratic. On any view a duty holder has a responsibility in regard to the safety of the services and products which it uses. There is also some force in the view that it is undesirable that an assessment of the adequacy of a safety case should become involved in an examination of the adequacy and safety implications of individual contractual arrangements. The evidence before me has not persuaded me that the safety case regime should be extended.

7.73   However, I consider there would be merit in a system by which companies supplying products or services for use on, or in regard to, railways in Great Britain, at least where such products or services are of a safety-critical kind, should be required to hold an accreditation as a condition of being able to engage in that activity. Accordingly I am in favour of the principle of such a system. The working out of the main features of such a system, let alone the details, plainly requires further study in order to determine, amongst other things:

(i)   what should be the definition of the categories of companies which would be subject to the system;

(ii)   what should be done in order to merit accreditation; and

(iii)   how accreditation is to be kept under review after it is granted.

As Miss Jenny Bacon, then Director-General of the HSE, observed in the course of her evidence:

"The one thing I would say is that, from HSE's experience, simply having accreditation does not necessarily guarantee competence. It has got to be a living matter and it has got to go on being checked by the people that are letting the contracts".

It would also be necessary to spell out, at least in guidance, what use may and may not be made of accreditation. While a railway operator may properly rely on the fact that it has obtained services from an accredited supplier, this should not be regarded as discharging its whole obligation in regard to the safety of what is supplied. Lastly, there is the question of the body by whom such a system of accreditation should be administered.

7.74   As regards a national system of licensing of individuals, which is a subject which went beyond the scope of the subjects which the Inquiry set out to investigate, I can at least say that I agree that there should be a system for the central licensing and recording of qualified drivers. This would cover their knowledge of the rules and regulations, including the Rule Book in so far as it is concerned with drivers and the traction for which they had been assessed as competent. There does not seem to me to be a need for centralised training for this purpose. It would suffice if training providers or train operators were accredited and common standards laid down. Drivers' licences should require to be re-validated every three years. As regards competence in local operational knowledge, on the other hand, I am not satisfied that licensing in this respect is a practical proposition. Each TOC should maintain its own record of the extent to which its drivers have been assessed as competent in regard to such knowledge.

7.75   I am likewise in favour of the central licensing and recording of qualified signalmen, based on the assessment of their knowledge of the rules and regulations, including the Rule Book in so far as it is concerned with signalmen. What I have said in the last paragraph about re-validation should apply here also.

Chapter 8
Railtrack and Railway Safety

Introduction

8.1   This chapter is concerned with the role of Railtrack in the safety case regime, and the functions performed by Railway Safety in regard to that regime and Railway Group Standards. In particular I will consider whether it is appropriate that Railtrack should have what was described by some parties as the role of a "quasi-regulator"; and whether the functions of Railway Safety should be taken over or superseded by the creation of a new body independent of Railtrack Group plc and their subsidiaries. This chapter should be read in conjunction with Chapter 9.

8.2   This chapter covers the following subjects:

- events leading up to the present position (paras 8.3-8.15);
- what was in issue at the Inquiry (paras 8.16-8.25); and
- discussion and conclusions (paras 8.26-8.38).

Events leading up to the present position

"Ensuring Safety on Britain's Railways"

8.3   The position which Railtrack occupy as "infrastructure controller" has its origin in the report by the HSC "Ensuring Safety on Britain's Railways", which they submitted to the Government in January 1993.

8.4   One of the principles which was regarded as fundamental in that report was, in the words of para 8b of the Executive Summary:

> "The prime responsibility for ensuring safety on the railway must rest with the party (or parties) who has (have) control (but this responsibility is limited to the extent that they actually have, or ought to exercise, that control)".

In discussing the basic requirements for a safety regime the report observed at para 4.1:

> "For a potential new operator running a service on the national infrastructure, or operating a station, that operator must be held accountable for those aspects for which he has control. But because that operator might be introducing risk into the railway environment which could affect the safety of the infrastructure and other operators' activities there must also be a responsibility upon the party who has control of the system itself to impose conditions on access (especially those related to safety) and monitor (by inspection, audit etc) what is going on".

The report went on to remark at para 4.18:

"Indeed because an infrastructure controller has practical control of access and movement on the system (not least because of having control of the signalling) it is essential that the control is exercised in a way which ensures safety 'so far as is reasonably practicable'. This cannot relieve new operators of their own responsibilities. Neither does it imply that the infrastructure controller is being given, or is taking on, an overtly 'regulatory' role. It is simply a question of appropriate arrangements to satisfy the infrastructure controller's own obligations".

8.5 On this basis the HSC put forward their proposal for a safety case regime, in which the "infrastructure controller" not only would produce its own safety case for acceptance by the HSE but also, under a "cascade", would have the function of accepting the safety cases produced by train operators. In this chapter I will, for brevity, omit reference to station operators, but the same applied to them.

8.6 Two points may be noted at this stage. They were not stated to be of decisive importance but were part of the context in which the HSE made their proposals. The first was that it was expected that "in the medium term" the whole of British Rail's infrastructure would be controlled by Railtrack as a public sector organisation. According to the evidence of Mr V P Coleman, Chief Inspector of Railways for the HMRI, this meant at least until the first round of franchising had been completed. Secondly, it is evident that the report took a conservative approach to change. The cascade system came into being at a time of extremely rapid change in the industry when new operators were being admitted to the national rail network. It is significant that one of the other fundamental principles which was adopted in the report included the words that

"…any system which emerges must not lead to any diminution of current safety standards".

8.7 The 1994 Regulations, following the recommendation of the HSC, conferred on the infrastructure controller, which for present purposes is Railtrack, the responsibility of accepting the safety cases of train operators, in accordance with the "cascade". This was subject to a right of appeal under Regulation 25 to the Secretary of State.

8.8 As regards railway standards, the report by the HSC stated, at para 19 of the Executive Summary:

"It is acknowledged that the infrastructure controller will have a co-ordinating role in relation to the application of new or revised standards which affect safety on the system, especially those affecting basic compatibilities and those at interfaces between parties".

The report went on in para 20 to state:

"In the longer term it will be appropriate to consider the development of the railways standards body (ies) into a position independent of BR. Given the present uncertainties about the future shape of the rail industry and about the likely views of new companies once they become established, it is not appropriate to plan now for the longer term future. It is recommended that a future study on this is undertaken some 4 years after the start of franchise operations on the network".

In the event the S&SD, who were one of Railtrack's directorates, took over functions in relation to standards, as I have described in paras 3.10-3.11.

The Rowlands Report

8.9   The outcome of the Rail Safety Policy Review, which was set up in the aftermath of the crash at Ladbroke Grove, was the Rowlands Report, which was published in February 2000. It recommended that the HSE should take over from Railtrack the responsibility for accepting the safety cases of train operators. The report did not give explicit reasons for this recommendation – which meant the ending of the "cascade" – but it is evident that this was to provide the public with assurance that the acceptance process was in the control of the independent safety regulator.

8.10   It may be noted that para 29 of the report stated:

> "On operator safety cases, Railtrack Line as the network operator is best placed to consider the risks imported into the network. Indeed it must continue to do this in order to fulfil its duties under (the 1974 Act). Railtrack Line should check operators' safety cases for interface and interworking issues and thus take responsibility for ensuring that operators' safety cases are sound. It should give its endorsement (or otherwise) to operators' safety cases. A safety case should not be accepted unless Railtrack Line is content, but we do not believe that this should be the only endorsement necessary".

The report went on, in para 30, to recommend that Railway Safety

> "…should be involved as an independent party in the consideration of individual operators' safety cases".

They should make an advisory recommendation to the HSE on whether they were fully satisfied that the safety case holder would not import undue risk on to the network. They should also advise on the issues to be considered in operators' safety cases and indicate the level of detail needed to meet regulatory and other requirements.

8.11   The report recommended the establishment of Railway Safety as a "Railtrack subsidiary" with a board comprising a range of industry representatives and independent persons. The report recommended that Railway Safety should have responsibility for the setting of Group Standards (including standards for new equipment) and for developing the intra-industry system of safety assurance. They should publish the Railway Group Safety Plan, which would include a programme for safety-related research. The reasons given for the recommendation that Railway Safety should be set up were to increase the independence of the activities performed by the S&SD and to separate them from the commercial interests of Railtrack.

8.12   The Government accepted the Rowlands Report. When the Deputy Prime Minister announced this to the House of Commons on 23 February 2000 he stated that it set out an interim position, and that "the final fate" of the S&SD would be decided following the report of this Inquiry.

The Railways (Safety Case) Regulations 2000

8.13   At this stage it may be helpful if I summarise the position of Railtrack and Railway Safety with reference to the 2000 Regulations, which came into force on 31 December 2000. For the sake of clarity and brevity I will slightly paraphrase what is stated in the regulations. Railtrack are an "infrastructure controller" which is defined in Regulation 2(1) as "a person who controls railway infrastructure". Under Regulation 5 a safety case which has been prepared by a train operator is to be submitted to both the HSE and Railtrack. Railtrack are to procure an assessment of the safety case by an "assessment body", and obtain from it a report of the assessment including a recommendation as to whether, in short, it is satisfactory from the point of view of the performance of Railtrack's health and safety duties, along with, where acceptance is not recommended, the reasons for that "decision". Railtrack are to provide their own recommendation as to whether or not the safety case should be accepted and, if it should not, the reasons for that recommendation. Railtrack's recommendation is not binding on the HSE (differing from the recommendation in the Rowlands Report), but the HSE have to have regard to it and if they are differing from it they are to give reasons for that decision. Under Regulation 7 a revision of a train operator's safety case which "will render the safety case materially different from the version last accepted" is not to be made unless the HSE have accepted it. Railtrack are to provide their recommendation as to whether or not such a revision should be accepted and the reasons for that recommendation. Likewise, if the HSE differ from that recommendation, they are to give reasons for that decision. Under Regulation 9 Railtrack are to procure an assessment body to undertake at intervals of not more than 12 months an audit of their own operations arising from control of railway infrastructure, and the operations of any train operator in relation to infrastructure which is in the control of Railtrack. Railtrack are to send copies of the audit reports to the train operator, any other railway operator who may be affected by matters to which the report relates, and the HSE. Under Regulation 12 Railtrack are to take all reasonable steps to ensure that any train operator carries out its operations in conformity with its safety case, in so far as it concerns the performance of Railtrack's health and safety duties. Under Regulation 13 Railtrack are to report to the HSE any failure by a train operator to conform with its safety case which is likely to increase the risk of serious personal injury; and any failure to comply with a reasonable request which Railtrack have made in regard to the performance of its health and safety duties.

8.14   Railway Safety are a body which is intended to meet the definition of an "assessment body". This is defined in Regulation 2(1) as meaning a body which, in relation to an assessment or audit to which I have referred in the previous paragraph, is:

"(a)   competent to carry out the assessment or audit, and

(b)   sufficiently independent of the infrastructure controller to ensure that the assessment or audit is objective".

In addition to carrying out an assessment of the safety case of a train operator, Railway Safety are also to carry out an assessment of Railtrack's own safety case, in terms of Regulation 4.

The establishment of Railway Safety

8.15   Complementing the changes to which I have already referred, the Rail Regulator modified Condition 3 (now Condition 6) of Railtrack's network licence so as to require the establishment of what came to be Railway Safety as a new subsidiary company of Railtrack Group plc. Para 3.2 of his Notice of Modification stated that this was in order to meet the following objectives:

- in parallel with the HSE's proposed modifications to the Railway (Safety Case) Regulations 2000, to strengthen the arrangements which ensure that the commercial interests of Railtrack do not impinge on the safety responsibilities currently discharged by the S&SD;
- to ensure that, having primary regard to safety requirements, the S&SD treats all industry parties in a fair and even-handed way; and
- to make much clearer the arrangements for separation of the S&SD from the functions of Railtrack plc and Railtrack Group plc.

In para 3.7 the Rail Regulator pointed that he had made it clear in a previous consultation document that he saw the changes as interim steps pending the outcome of the present Inquiry. According to the modified terms of Condition 3, the functions of the intended body included the provision of advice to the HSE, the Rail Regulator, Railtrack and stakeholders in respect of matters relating to safety of the operation and use of Railtrack's network; the use of the licence holder's network; various functions in relation to the Railway Group Standards Code; the discharge of any statutory obligation in relation to safety imposed on or in respect of any person carrying out the IRSA; and the promotion of research and development, training and the provision of information relating to all aspects of safety relevant to railway services on or relating to Railtrack's network.

What was in issue at the Inquiry

The parties' positions

8.16   At the end of Part 2 the parties, with one exception, supported or, in the case of Railtrack accepted, the transfer from Railtrack to the HSE of the function of accepting the safety cases of train operators, in accordance with the 2000 Regulations. For brevity I will omit reference to the acceptance of revisions, to which the same applies. The exception was the RIA. In their statement of case – which was not enlarged upon in any closing submission – they stated that they believed that the "cascade" was sound and suitable for the situation in which Railtrack were in private ownership and for the management of safety cases. Railtrack, who had previously expressed objection to removal of the "cascade", an objection which was strongly supported by the evidence of Mr R I Muttram, Director of the S&SD, stated in their closing submission that, while the principles which had been laid down by the Robens

Committee in 1972 led, in their view, to the conclusion that such safety cases should be accepted "at industry level", it was recognised that

> "…there has been a substantial loss of public confidence in the structures of the railway industry and its regulators such that some change is probably essential. Such modification should not be radical and should go no further than the 2000 Regulations".

8.17   The other changes which had been brought about following the Rowlands Report had few supporters in the Inquiry.   All of the parties, with the exception of Railtrack, Amey Rail, EWS and the RIA, expressed the view that the functions of Railway Safety should be taken over, or superseded, by a body which was independent of Railtrack Group plc and their subsidiaries, either a new rail safety authority or a new rail standards authority or body (see Appendix 5).   It may be noted that, beyond welcoming the transfer of functions from the S&SD to Railway Safety, Amey Rail expressed no view.   As regards the RIA, the Inquiry was informed that their members were divided as to whether Railway Safety should remain within Railtrack's corporate structure or should have greater independence.   Railtrack pointed out that the report by the HSE in September 1999 "Review of Arrangements for Standard Setting and Application on the Main Railway Network" (referred to as the Tansley Report) stated that there did not seem to be any cause for immediate concern on safety grounds in the way that the S&SD had operated their key safety functions.   The report by the HSE in February 2000 "The Management of Safety in Railtrack" found that safety and commercial functions were diligently separated.   Railtrack also said that they set store by the fact that there never had been an appeal against a standard which had been set by the S&SD.   However, Railtrack accepted the changes which had led to the establishment of Railway Safety, and advocated that they should retain their functions.

Railtrack's arguments

8.18   At this stage it is appropriate that I should set out in some detail Railtrack's arguments for the retention of that status quo, which were spoken to in evidence by Mr Muttram. The sheet anchor of their case was their claim that they had responsibility for control of the importation of risk to the infrastructure, which had been recognised at the outset in "Ensuring Safety on Britain's Railways".   This responsibility was based on their interpretation of their duties under Sections 2-4 of the 1974 Act.   Particular emphasis was laid on their duty under Section 4, as a person having control of "premises", namely the infrastructure,

> "…to take such measures as it is reasonable for a person in his position to take to ensure, so far as is reasonably practicable, that the premises…are safe and without risks to health".

Railtrack had exercised that control by:

(i)   mandatory standards for access to the infrastructure, including the acceptance of new vehicles;

(ii)   the exercise of functions under the regulations relating to the safety cases of train operators; and

(iii) their contractual relationship with train operators (through track access conditions) and with contractors working on their behalf (through conditions of contract).

8.19 As regards Group Standards, Mr Muttram said that they were "Railtrack's standards" for the purpose of securing safe systems and safe interworking. He added:

"We do, of course, operate the participative and consultative processes in the way we produce those standards to try to make them as fair to the industry and as pan-industry as we can".

When the 2000 Regulations came into force these standards would remain Railtrack's. Railtrack were responsible for ensuring that they were sufficient. He accepted that how the cost imposed by the need to comply with Group Standards fell as between them and train operators was subject to the Railway Group Standards Code, in Railtrack's control.

8.20 As for safety cases, Mr Muttram made it clear, as I have already stated, that he called in question the loss of the "cascade". He also said that the complex interaction between trains and the infrastructure, along with the operation of both, created, in his view, the need to secure that train operators had systems in place which were capable of ensuring that the risks which their operations introduced or might introduce to the network were properly controlled. He referred to this as "an assurance activity, it is not a control activity". As regards contractual relationships, he accepted that under a standard condition of track access agreements, Railtrack had the right to suspend access by a train operator to part of the network in the event of a breach which they reasonably considered to constitute a serious threat to the safe operation of the network. However, this was "a nuclear option", and there was room for dispute as to whether it was reasonable for Railtrack to take this step. Railtrack maintained that there was a need for both Railtrack and Railway Safety to have graduated powers, especially if, as the HSE maintained, the appropriate course for Railtrack to take in order to deal with non-compliance was to exercise a contractual remedy. Such powers should enable action to be taken which was "appropriate to the exigencies of the situation".

8.21 Railtrack emphasised that control should not be divorced from power. Only one person should be primarily responsible for the network. Mr Muttram said that if the body in charge of Group Standards was not Railtrack,

"...you effectively have to accept that Railtrack is no longer the premises controller under the '74 Act, and that means it is not responsible in any way for the conduct of people on its infrastructure. That seems to me to be diluting safety, not enhancing it".

Railtrack also pointed out that Miss Jenny Bacon, then Director-General of the HSE, referred to Group Standards as

"...generic contractual terms rather than regulations, in the sense that I understand regulations at any rate".

She went on to say:

> "I do think it is important that Railtrack should have primacy in setting those generic contractual terms, because of the point that has been made repeatedly about the importation of risk on to the railway".

Railtrack stressed the importance of not compromising their obligation to secure co-operation from other railway operators.

8.22   Mr Muttram also gave evidence that even if an external body discharged the functions of Railway Safety, Railtrack would still need to satisfy themselves as to what was adequate. This would involve duplication of resources, in addition to the recruitment for themselves which they had already had to undertake in order to deal with the change from the S&SD to Railway Safety.

8.23   As regards Railway Safety, Railtrack submitted in closing that Railtrack Group plc

> "...should remain the dominant guarantor of Railway Safety because Railtrack is the infrastructure controller. Railtrack Group needs, via Railway Safety and Railtrack, to retain ultimate control over the interface between wheel and rail and other related safe interworking issues. This dominance is balanced by ensuring that Railway Safety Board Members are drawn from the industry as a whole...and by the licence controls over Railtrack. Thus sufficient independence and objectivity of Railway Safety's actions is transparently achieved and the regulation of Railtrack's licence precludes abuse of its position".

Railway Safety, it was said, would derive no powers from Railtrack or the HSE or statute, but would simply be an independent advisor. Railtrack contended that it was of overriding importance that standards of the type of Railway Group Standards

> "...should be set by the industry for the industry and not prescribed by a safety regulator".

8.24   Mr Muttram said that, bearing in mind the complexity and variability of the network, which had been built over a very long time, it would be very difficult for an external body to carry out the necessary checks. It would require an education process, a dialogue, going on all the time to bring it "up to speed". For it to deal with derogations would be a "bureaucratic nightmare". Railtrack expressed the concern that any such change might delay the development of standards which were essential to realising major railway projects.

8.25   Mr Muttram went on to say that there was a risk of conflict between external power and internal control. If Railtrack considered that a Group Standard which had been set by an external body was inadequate, they would have the right, unless it was taken away by law, to insist on an overriding standard which gave greater control of risk. He was also concerned as to the question of liability if there was a disagreement of that kind as to what was or was not acceptable.

Discussion and conclusions

8.26   The pre-Rowlands regime plainly accorded a dominant position to Railtrack. Under the 1994 Regulations, as the "infrastructure controller", they were given the function of deciding – admittedly subject to appeal – whether the safety case of a train operator should be accepted or not. By the Railway Group Standards – which they regarded as their standards – they set the conditions on which train operators were allowed to use their network. Railtrack did not accept their role was that of a "quasi-regulator", but that description was justified since they clearly exercised control through the rules and restrictions which they imposed.

8.27   In my view it was inappropriate that a commercial organisation, such as Railtrack became in 1996, should continue to fulfil such a role in relation to other commercial organisations such as train operators. As ATOC submitted, this was the superimposition of a relationship of regulator and regulated upon a relationship of supplier and customer. It carried with it real or perceived risk of conflict of interest between Railtrack's commercial interests and their responsibilities under the Safety Case Regulations and the system for the setting of Group Standards. ATOC cited as an example of potential conflict the requirement that had been made by the S&SD in April 2000 that train operators should fit sanders to all their trains. They pointed out that problems with low adhesion were commercially disadvantageous to Railtrack.

8.28   In view of these considerations, which I find compelling, I endorse the transfer from Railtrack to the safety regulator of the function of acceptance of the safety cases of train operators and the removal from the S&SD of their functions in regard to safety cases and Group Standards. The question then is whether the process of reformation has gone far enough. As I have already pointed out, although the "cascade" has been brought to an end, Railtrack still have the function of making recommendations, and the HSE have to give reasons if they are to differ. Railtrack also supply the HSE with a report by Railway Safety of their assessment. Railway Safety are not a subsidiary of Railtrack and hence, as ATOC pointed out, they are not legally bound to act in accordance with the wishes of Railtrack. On the other hand Railtrack insist that it is necessary for both them and Railway Safety to remain under the control of their common parent company. It is clear that Railtrack are still accorded, and seek to maintain, a dominant position. The supplementary submission by ADtranz, which accompanied the statement of case for the RIA, is of some interest at this point. At paras 6-7 ADtranz remarked, in the context of the introduction of new rolling stock,

> "The potential for commercial conflicts of interest and for safety judgements to be coloured by commercial considerations is obvious and ADtranz recognises that the proposed structure of Railway Safety has been fixed upon in an attempt to address these concerns.
>
> The fact remains that Railway Safety as currently proposed would perpetuate an arrangement within which funding, staff, promotion prospects and hence general organisation culture would flow mainly from Railtrack regardless of the composition of the Railway Safety board. The wider composition of the Railway Safety board would give Railway Safety an appearance of independence that would not in fact be borne out at working level".

8.29   In commercial terms Railtrack are plainly a supplier, that is to say a supplier of access to their network. As I have already noted, Railtrack laid great stress on the contention that their role as controller was a consequence of their duties under the 1974 Act, and in particular Section 4. At the same time, as was pointed out, it is important to remember that the duty under that section contains a double qualification. It is limited to what is reasonably practicable, and to whatever measures are reasonable for a person in the position of the duty holder to take. The passage from the evidence of Mr Muttram which I quoted in para 8.21 should be taken along with the statement made by Counsel for Railtrack in his opening submission at the outset of Part 1 of this Inquiry:

> "Railtrack is widely but erroneously seen to be responsible for ensuring safety on the whole rail network and to have the power to do so".

I do not suggest that the duties of Railtrack under the 1974 Act do not lead to Railtrack having an important interest in ensuring that what they operate, namely the infrastructure, is and remains safe. However, as Counsel for the Rail Regulator said in his closing submission,

> "…the rail system necessarily involves risk. Having trains involves risk: operating lines, signals and stations involves risk: carrying out repair and maintenance involves risk. All the constituent parts of the industry have to work together to minimise both the risk each of them creates and the risks arising from their interaction".

8.30   This brings me to a submission made by ATOC that "Ensuring Safety on Britain's Railways" and the safety case legislation have been based on an erroneous assumption that risk flowed only one way. ATOC cited as an example the previous attitude which had been taken to SPADs as being a matter of driver error. An unacceptable risk could be caused by the way in which the infrastructure was managed and maintained. There is considerable force in this criticism of the orientation of the regime. As Counsel to the Inquiry pointed out, in the event of a crash the persons most likely to be injured are employees of a train operator and its passengers, as opposed to those who were employed by Railtrack or working under contract with them. Counsel for ATOC pointed out that there was no machinery by which train operators could control the exportation of risk by the infrastructure to their trains, and there was no explanation as to why this was so. The "importation of risk" principle had led to an "asymmetry" in the regulatory regime which was entirely different from any other transport industry regime. As regards the 2000 Regulations, it was pointed out that, while Railtrack were to provide the HSE with their recommendation, no provision had been made for train operators to be given the same opportunity in regard to Railtrack's safety case. Mr Waite agreed that the regulations enabled Railtrack to exercise "a major influence on the HSE's decisions". He expressed the view that there would be benefit in having more of a two-way flow of information between Railtrack and the train operators, although he did not think that they needed to be given a formal opportunity to review Railtrack's safety case and satisfy themselves as to its adequacy. When asked about the question of equality of treatment, Mr Coleman said to the Inquiry that although the 2000 Regulations did not specifically mention taking into account the view of other duty holders, that did not mean that the HSE would not do so. He accepted, however, that there was a question of public perception. The introduction of a requirement that copies of the audit on Railtrack should be more widely available was welcomed by ATOC, particularly in view of difficulties which train operators had experienced in attempting to obtain copies of the audits of Railtrack Zones, despite assurances which had been given by senior officials. However witnesses gave evidence that the provisions as to auditing in the 2000 Regulations did not go far enough, and that it should be open to any railway operator to call for independent audit of any aspect which was causing concern.

8.31 Mr Muttram said he did not agree that there should not be asymmetry. This was

"…because Railtrack actually controls the network and it is the network that actually makes the complexity".

He said that train operators would not necessarily have the same overview as Railtrack, but he accepted that much more information should be passing from Railtrack to the train operators. The latter had the right to be assured that Railtrack were doing their job properly as well.

8.32 In my view the time has come for more radical alterations than those proposed by the Rowlands Report. To a significant degree its recommendations have proved to be an unsatisfactory half-way house. It is, of course, fully understandable that in the circumstances it was thought appropriate to advance only a certain distance, pending the outcome of the present Inquiry.

8.33 It is and remains correct, in my view, to regard Railtrack as having control of "premises", namely the railway infrastructure of their network. Thus they are for the purposes of the Safety Case Regulations an "infrastructure controller". They have an important interest in regard to whether the safety case of a train operator or a material revision of such a safety case is accepted or not, since how the train operator manages the risks which are its concern, in conjunction with the activities of other railway operators, may affect the question whether Railtrack comply with their own duties under health and safety legislation. However, the provision that the safety regulator is to look to Railtrack for a recommendation indicates a continuing dominance on the part of Railtrack which is not justified. If the question of acceptance of the safety case of a train operator, or its revision, is one for the safety regulator, as it should be, it should not be pre-judged by one of the industry parties, no matter how important is its interest. Instead the safety regulator should give Railtrack the opportunity to make any representation as to whether or not the safety case or revision should be accepted, and the grounds on which such a representation is based. The safety regulator should likewise give the opportunity to any other train operator which may be affected by matters referred to in the safety case to make a similar representation, and for this purpose select whichever train operators it considers to be appropriate in the circumstances. If the safety regulator refuses to accept the safety case or its revision, it should give the reasons for that decision.

8.34 As regards the safety case for Railtrack or any material revision of that safety case, I consider that a similar approach should be taken. Accordingly, to the extent to which it considers it appropriate in the circumstances, the safety regulator should give any train operator the opportunity to make a representation as to whether or not the safety case or revision should be accepted, and the grounds on which the representation is based.

8.35　As to Railway Safety, an attempt has been made to secure an acceptable degree of independence. The integrity of those who administer the affairs of Railway Safety (or their predecessor) is not in question. However, it is plain that the relationship between Railtrack and Railway Safety also shows an attempt to achieve two incompatible objectives. It is clearly intended that Railtrack Group plc should control via Railtrack and Railway Safety "the interface between wheel and rail and other related safe interworking issues". This inevitably seems to draw Railtrack and Railway Safety together as part of a single means of exercising that control. Another somewhat contradictory position relates to Group Standards. From time to time they were described by Railtrack as standards set by the industry at industry level. But this is in order to distinguish them from standards set by some external body. They are plainly regarded by Railtrack as their standards, setting out the basis upon which, as far as they are concerned, others are allowed to operate on their network.

8.36　Railway Safety have been given the role of fulfilling functions required by the 2000 Safety Case Regulations. In considering whether a safety case should be accepted the HSE are apparently dependent to some extent on an assessment made by Railway Safety. Railway Safety cannot be regarded as fully independent of Railtrack Group plc and their subsidiaries. In my view the safety regulator should be wholly in charge of the assessment of a safety case. It should be for the safety regulator to decide to what extent, if at all, it should commission assessment from an independent body for that purpose. In these circumstances Railway Safety should cease to discharge the function of assessment for the purposes of the Safety Case Regulations. It may well be that members of the staff who presently carry out this work can be subsumed within the Railtrack organisation.

8.37　I am in favour of there being a provision in the Safety Case Regulations which would impose a duty on Railtrack to carry out, or procure the carrying out by a suitably qualified body of, audits for the purposes presently set out in Regulation 9 of the 2000 Regulations. By this means Railtrack would examine the detail of compliance at first hand. The safety regulator, on the other hand, should be in overall charge, by, for example, reviewing the adequacy of Railtrack's auditing, carrying out its own audits to the extent that it considered appropriate, and dealing with instances of non-compliance whenever they arose. I am not persuaded that it is necessary to confer additional powers on Railtrack, especially in view of the escalation procedure and their ability to rely on having made a reasonable request under Regulation 11 of the 2000 Regulations. Regulations 12 and 13 of the 2000 Regulations are also of significance and should remain in effect (cf para 7.41).

8.38　As regards the setting of Railway Group Standards, I consider that it is important that they are, and are seen to be, set independently of the control of any one member of the industry. If any other body than Railway Safety is to assume that function it will need to command the confidence of all, including Railtrack. It will have to be fully aware of all the issues relating to the interface between wheel and rail and safe interworking. It will have to take full account of the knowledge, views and interests of the body of railway operators, including Railtrack as its principal member. While such a change would not mean that Railtrack would lose their right to insist upon a higher standard than that set by an independent body, the occasion for this seems to me to be theoretical rather than real. I see no good reason why an independent body should not enjoy the confidence of the industry, or that its relationship with Railtrack should be a scene for conflict. In my view an independent body should be well able to carry out the functions with competence, efficiency and fairness. It will be for Railtrack to ensure that their Assurance and Safety Directorate are adequately equipped for the purpose of supporting their own interests. This function should be assumed by a body which is independent of Railtrack Group plc and their subsidiaries. In Chapter 9 I consider whether the safety regulator should be that body.

Chapter 9
The safety regulator

Introduction

9.1   This chapter is concerned with the role of the safety regulator, the functions which it should discharge, and its relationship with the Rail Regulator and the SRA. It takes up the question of the functions presently discharged by Railway Safety as part of a discussion of a number of models for a new safety regulator which were proposed by parties to the Inquiry.

9.2   The chapter will cover the following subjects:

- matters of concern (paras 9.3-9.6);
- the role of the safety regulator (paras 9.7-9.18);
- the responsibility for accident investigation (paras 9.19-9.30);
- a new safety regulator?:
  - the models (paras 9.31-9.38);
  - the responsibility for Group Standards (paras 9.39-9.46);
  - the choice of safety regulator (paras 9.47-9.60);
  - conclusion (paras 9.61-9.66);
  - European regulation (paras 9.67-9.74);
- the safety regulator and the public (paras 9.75-9.80);
- the relationship between the safety regulator and the Rail Regulator (paras 9.81-9.97); and
- the relationship between the safety regulator and the SRA (paras 9.98-9.112).

Matters of concern

9.3   A recurring theme during the discussion in Part 2 of the Inquiry was the need for simplicity and clarity in the structural arrangements for the delivery and regulation of safety in the rail industry. The present arrangements were described as containing room for confusion, uncertainty and duplication. In particular this related to:

(i)   the definition of the role and responsibilities of different bodies, and their inter-relationship; and

(ii)   the arrangements for the auditing of compliance with Group Standards and with health and safety legislation.

These are matters of concern not merely to those who work in the rail industry or who seek to regulate it. They have to do with the level of public confidence in the way in which safety on the railways is managed and regulated.

9.4 Along with these concerns are questions as to whether the regulation of safety should be extended to cover areas which are not at present covered, for example whether there should be a system for the licensing of suppliers, equipment and personnel. Another important question is whether the safety regulator should assume responsibility for Group Standards. Another is whether it should seek to apply a more prescriptive approach to safety regulations than has been done in the past. These and other concerns are sharpened by a sense that there have been significant weaknesses in the culture and management of safety in the rail industry.

9.5 The position of the HSE as the present safety regulator has itself been called in question. It has been proposed that because of their past weaknesses and failures they should be superseded by a new safety regulator specifically for the rail industry.

9.6 No doubt if there were easy answers to these questions they would have been found some time ago. It is plain that the undoubted complexity of the structural arrangements springs to some extent from the disaggregated state of the rail industry. I have to take the constitution of the industry as it exists. It seems to me that in considering structural change the most productive approach is to ask what can be done to enable safety to be as effectively managed and regulated as it would be if the industry were a single enterprise.

The role of the safety regulator

9.7 Having in mind some of the issues which arise in this chapter, I will make a number of general comments about the role of the safety regulator. The discussion in Part 2 clearly showed general agreement that there should be a single body which has responsibility for regulating safety on the railways. That has, of course, been the position from the outset of privatisation. Thus one of the fundamental principles which was stated in "Ensuring Safety on Britain's Railways" was that

"…legislation pertaining to railway safety should be administered by a single independent safety regulator, the HSE".

9.8 Next, it was generally agreed in this Inquiry that, with regard to the regulation of safety on the railways, there should not be any distinction between the safety of the public and the safety of those who were employed to work on them. The two are inevitably intertwined, and it would make no sense to have separate regulators for each.

9.9 There was unanimity that it was essential that the safety regulator should have adequate strength. Mr J Rimington, formerly Director-General of the HSE, put the matter succinctly when he wrote in his submission to the Inquiry of

"…a unified regulatory body well resourced and powerful enough genuinely to challenge vested industrial interests and stimulate change".

9.10 It is plainly necessary when considering the role of the safety regulator to look at it, not in isolation, but in its relationship to the responsibility of the members of the rail industry. Another of the fundamental principles of "Ensuring Safety on Britain's Railways" was that

> "…the prime responsibility for ensuring safety on the railway must rest with the party (or parties) who has (have) control (but this responsibility is limited to the extent that they actually have, or ought to exercise, that control)".

The Rail Regulator observed at para 34 of his statement of case:

> "There should be no dilution of the principle that it is for the industry itself to protect the safety of those who use and work on the railway".

In his evidence Mr Tunnicliffe, formerly Chief Executive, London Transport and Chairman, LUL, enthusiastically described that principle as "an absolutely brilliant approach to safety", remarking that the industry was well served by the 1974 Act.

9.11   The overall arrangement was succinctly described by Mr M J Beswick, Director of Network Regulation for the ORR, when he stated:

> "The role of a safety regulator is to monitor and inspect the processes which are in place, proactively to challenge and to enforce, and to create a climate conducive to safety improvement, innovation and practice, rather than deliver safety. He should monitor and approve industry safety arrangements".

9.12   At the same time it has to be recognised that in the disaggregated state of the rail industry there are a number of essential functions which are beyond what the management of safety by an individual operator can achieve. The most obvious example is the setting of Group Standards for safe interworking and the auditing of compliance with those standards by all to whom they apply. One of the matters which I will require to consider in the course of this chapter is what is the most appropriate way of accommodating these functions. In the course of his evidence Mr A R Taig, Managing Partner of Risk Solutions, which is a subsidiary of AEA Technology plc, said there were two fundamental requirements to ensure the safety of the railways, or any other transport system. These were:

> "(i)   Each individual element of the systems must play its part towards delivering railway safety, and
>
> (ii)   the elements must fit together so that when each plays its part properly, they collectively deliver the whole system safety we require".

9.13   Mr Taig went on to state, in the course of his useful analysis, that:

> "…only a publicly accountable regulator can provide the overall policy framework, determine what safety outcomes society wishes from the railways, and how these fit alongside other policy goals".

This included the question of what degree of risk was tolerable and how incremental changes in risk should be valued. On the other hand, the function of "system architect", i.e. devising the best way to satisfy those policy outcomes, might or might not be carried out by the safety regulator.

9.14   Mr Taig went on to say that, in his view, the industry

> "…is best placed to have ownership of the design of its arrangements (standards rules, etc) and also of the front line assurance processes by which they make sure the arrangements are effectively applied",

whereas

> "…the regulator must satisfy itself that these arrangements are effective, and have the power to intervene if they are not – either in the devising of improved standards, rules and procedures, or in the inspection, assurance and other compliance processes".

9.15   For the moment I express no view about the division of responsibilities which was described by Mr Taig. However, it is clear that on any view, a line has to be drawn between what is regarded and treated as matters which should be left to the industry and those which are properly for the safety regulator.

9.16   If the safety regulator is to discharge its function properly and give the public notice that it is doing so, it is essential that it should be independent of the industry and be clearly seen to be independent of it.

9.17   Throughout it has to be recognised that, while the management and regulation of safety, including the application of the lessons learned from accidents and incidents, are highly important elements in preventing accidents, absolute safety can never be guaranteed, no matter what the cost.

9.18   One of the basic principles of the approach of the Robens Committee in 1972, which led to the enactment of the 1974 Act, was the aim of minimising the statutory regulation of safety. Thus another of the principles which was regarded as fundamental in "Ensuring Safety on Britain's Railways" was that

> "…the degree of statutory control shall be the minimum consistent with the need to ensure adequate and cost effective levels of control of risk and to secure public confidence".

It is not in doubt that there is, and should be, room for both goal-setting and more prescriptive types of regulation. They are not exclusive of each other, but they represent different areas within a single spectrum. However, it is important to bear in mind that considerations of policy may influence a shift in one direction or the other, either in general or in respect of some specific subject. In the course of her evidence Miss Jenny Bacon, then Director-General of the HSE, said:

> "When the safety culture and risk management in an industry are weak, prescriptive regulation may be a necessary step on the road to improvement".

The responsibility for accident investigation

9.19 Before coming to the case for a new safety regulator it is convenient to deal with the question whether the responsibility for accident investigation should or should not continue to be discharged by the safety regulator, whether that regulator is the HSE or is to be some other body. Under the Reporting of Injuries, Diseases and Dangerous Occurrences Regulations 1995, all accidents and incidents have to be reported to the HSE. A decision is made as to whether there should be an investigation or an inquiry under Section 14(2)(a) or (b) of the 1974 Act. In the case of the rail industry this is in addition to a formal inquiry or formal investigation carried out under Group Standard GO/RT 3434/3, which replaced the British Rail standard in June 1997. According to an internal HMRI document of 1999, their investigations serve a variety of purposes including

> "• ensuring that the circumstances of the particular incident are identified and rectified, and identifying any breaches of legislation;
> • assessing the quality of standards and safety management generally (particularly in relation to the railway operator's safety case) and motivating management generally towards improved health and safety standards etc; and
> • providing information to assist HMRI/HSE to formulate policy, guidance, standards etc, and also to satisfy expectations from the workforce, public, Ministers etc".

9.20 Mr D C T Eves, Deputy Director-General of the HSE, described the nature of the HMRI's investigations as

> "...technical, legal and advisory: technical in the sense of the need to understand and review the failures of industrial plant or management systems, legal because of the need to establish whether there had been a breach of law by duty-holders (the evidence-gathering process possibly leading to prosecutions), and advisory because of the need to disseminate any lessons amongst industry at large for the benefit of health and safety".

Proposals for change

9.21 Among the parties to the Inquiry there was what Counsel to the Inquiry described as "overwhelming" support for the setting up of an independent body for the investigation of railway accidents and incidents, hence relieving the HMRI of this responsibility. Thus, for example, ATOC proposed a RAIB, which would be similar to the AAIB and MAIB for air and marine accidents, and, like them, funded by the Government. I note that Recommendation 81 of Professor Uff in his report on the Southall rail crash was that consideration should be given to whether such an independent accident investigation body should be created. The proposal which was put before this Inquiry involved not only the supersession of the HMRI but also placing industry investigations under the overall control of the same body. I will refer to the part played by the industry later in this report. For the moment it is sufficient to note that it was envisaged that the RAIB would themselves investigate accidents and only the more serious incidents. Other incidents would be investigated by the industry under a form of delegation.

9.22   The principal argument which was advanced in favour of this proposal was that of structural conflict: it was inappropriate for the safety regulator to carry out the function of investigation since it might be necessary for the investigation to examine the decisions and activities of the safety regulator itself.   As the Rail Regulator observed in his statement of case:

> "…a safety investigator should be free, where necessary, to criticise the safety regulator if shortcomings on its part have contributed to the accident or its consequences. If the investigator and the regulator are one and the same, it may be difficult to convince the public that this aspect of the investigation will be pursued with the necessary vigour".

Other parties emphasised that the independent activity of the investigating body would provide a positive check on the functions performed by the safety regulator.

9.23   It may be noted that the consultation document issued by the Transport Safety Review (TSR) team of the DETR stated at para 2.22, when discussing the proposition of an independent cross-modal transport accident investigation body:

> "The reason for an accident may lie in flawed policy-making or in failings in either the setting or policing of safety standards. Accident investigators must not feel constrained in considering such possibilities".

In the response document which was published in June 2000, it was stated at para 3.17:

> "There is a view widely held amongst those concerned with aspects of transport safety that its regulation, and the investigation of transport accidents, should be kept separate. This was a view echoed in the response to the consultation, and underlined by very many of those with whom the Review Secretariat discussed the issues overseas. There was some acceptance that such a separation may mean a loss of synergy between the two activities, but this was not considered a decisive argument. The emphasis is laid on the accident investigator's freedom to conclude and to report that shortcomings in the regulatory regime contributed to the cause of an accident".

9.24   In this Inquiry Railtrack emphasised that there were a number of benefits in separating enforcement from investigation. These included the greater ability to focus on root causes and to identify the lessons without blame requiring to be apportioned; the development of enhanced investigatory skills and efficiency in process; and improvement in the efficiency with which lessons were disseminated.

9.25   It was pointed out that the briefing issued by the European Transport Safety Council (ETSC) recommended that the EU take steps to ensure that all railway accidents are investigated by independent bodies. I note that in a letter to the Inquiry dated 30 October 2000 Mr H Hilbrecht, Director of Directorate E: Land Transport, of the European Commission, indicated that a planned European Directive on Railway Safety was intended to deal with, *inter alia,*

"…the setting up of requirements that serious railway accidents and near misses must be thoroughly investigated by national bodies that are independent of railway undertakings, infrastructure managers and regulatory bodies".

The report by National Economic Research Associates (NERA) on Safety Regulations and Standards for European Railways, for DG Energy and Transport, pointed out that in the Netherlands, Sweden and Finland rail accidents were investigated by independent bodies constituted for that purpose. They were independent of the transport sectors which they investigated and also independent of the transport ministries and other regulatory bodies. The report also recorded that the experience in countries with such transport safety boards, which also included the United States, Canada and New Zealand, had been positive. It stated at para 4.3.3.2:

"Increasingly, it has become recognised that such boards must focus on the identification of systemic safety deficiencies, to the exclusion of placing blame for accidents. Achieving this requires the independence of the board, the ability to call on expertise as needed, and protection from court proceedings determining blame and punishment".

The report recommended the concept of independent accident investigation. The Inquiry heard evidence directly from witnesses who had safety responsibilities in regard to the railway systems in Canada and New Zealand. Speaking of the merits of independent accident investigation, Ms Faye Ackermans, General Manager of Safety and Regulatory Affairs for the Canadian Pacific Railway, said:

"To me the biggest disadvantage of having one agency with two halves is the lack of ability to really look in the mirror carefully".

Other criticisms of the existing system

9.26   A number of the parties expressed criticisms of the present system for accident investigation by the HMRI. The Joint Rail Unions were concerned about the relatively low number of investigations, but accepted that the HMRI had stated that, conditional on the outcome of their bid for additional resources, it was intended to increase the level of investigations. Counsel for the bereaved and injured represented by the Southall and Ladbroke Grove Solicitors' Group criticised the HMRI investigations for a lack of "transparency". They pointed out that reports of these investigations had not been published as a matter of course. The recommendations and the progress towards their implementation had not been published at all, save to the extent that they were referred to in annual reports of the HMRI. It was, however, noted that the HMRI recognised the need for change, and the paper by the HSE in June 1999 "Major Incident Response and Investigation Policy and Procedures" stated that in principle the main and internal reports should be published. The Collins Passengers' Group pointed out that there was no machinery for taking forward the recommendations of investigations or inquiries under Section 14 of the 1974 Act. The Group also drew attention to steps which the AAIB took to keep families informed throughout the investigation process, to give them the means of making contact at any time and to brief them on the circumstances and about any report before it was published. The AAIB also considered that important information could be obtained from the families. In this connection it may be noted that Mr K P R Smart, Chief

Inspector of Air Accidents in the AAIB, commended the form of aircraft passenger questionnaire which was used by the National Transportation Safety Board (NTSB) in the United States of America. While there was no suggestion that the HSE had not acted properly or without due consideration for the feelings of the bereaved and injured, the Group maintained that they did not have the ability or the resources to give the support which was needed. The emergency services were not trained to obtain information from passengers. It was noted that Mr Eves said that the HSE had been considering adopting some of the practices which were followed by the AAIB.

The position of the HMRI

9.27   The HSE responded that they accepted that it was necessary to demonstrate the independence of their investigations from their regulatory activities, that is to say inspection, acceptance of safety cases and approvals. Mr Eves explained the arrangements which had been put in place with this in mind. He said that there were two strands to the investigation, the first being the investigation of an accident and the second being the investigation of the HSE's prior involvement with the duty holder. A recent example of the latter was the internal inquiry into the crash at Ladbroke Grove, which was supervised by an Inquiry Board including an independent member. The incident investigation was overseen by a board, which included experts independent of the HSE, and was headed by a senior HSE manager who was not connected with the regulatory functions for that industry; and the investigation team was led by a manager who had no previous involvement with the duty holder. The inquiry into the HSE's prior role was modelled on similar lines. The Inquiry Board comprised senior staff of the HSE from outside the division responsible for the duty holder. In addition there would always be an independent expert on the board. The inquiry team, again from outside the duty holder's division, would undertake the analysis of the role of the HSE and reports their findings to the Inquiry Board. Counsel for the HSE obtained from Mr G R Profit, Group Director of Safety Regulation in the CAA, an acceptance that "hopefully it might work very well". On the other hand Mr Smart, when asked whether an arrangement which involved a "Chinese wall" within a single regulator or investigator would be adequate, responded: "I do not think that is likely to inspire confidence among the industry or the public".

9.28   Counsel for the HSE emphasised that there was no evidence that there had been any actual conflict between the regulatory and investigatory roles of the HSE. Strong reasons and a certainty of achieving advantages would be required before changing the present arrangements would be justified. The HSE also emphasised a number of factors. First, as Mr Eves pointed out, inspection and investigation techniques "feed off each other". In that connection it may be noted that the TSR team noted in their response report that it was important there should be sensible interaction between the activities of regulation and investigation, including a transparent and speedy process whereby lessons learned through an investigation could be fed back promptly into regulation (para 3.19). Mr Smart confirmed the importance of combining investigative techniques and knowledge of the subject. Mr I S Naish, Director of Investigations (Rail and Pipeline) for the Transportation Safety Board of Canada, acknowledged the danger of investigators

"...getting out of touch because the unfortunate thing is we are on site after a crash and that is not the normal day to day operations".

However, he went on to say:

> "But the point is we do have contacts in industry, we have contacts with the regulator, and it does not take very long once we get on site and talk to people to get a feel for what is going on. But we have to be a certain distance away because we are independent".

Mr I M Waldram, Past-President of the IOSH, observed:

> "All my experience tells me people who are good accident investigators are also good auditors and, therefore, it is very useful to be able to have a single organisation that has both of those activities".

The HSE also emphasised the many opportunities for cross-fertilisation of ideas and experiences which supported their policy work. Further, the HSE provided a highly integrated technical and forensic support available to their inspectorates. They had a wide range of interests and expertise, and could make very flexible arrangements. It was also emphasised that while the AAIB sought to ascertain the causes of an accident and the lessons which should be learned, investigations under the 1974 Act not only achieved these objectives but also brought home criminal liability. This contributed to the process of safety learning as well as enforcement. In the case of the investigation into the crash at Hatfield, the HSE had been able to field a wide range of expertise at short notice. It was both sensible and possible for the same body to conduct a technical investigation and to prosecute. It was suggested that the situation in aviation was not comparable. In that field there was no presumption of criminality and only a "light touch" in prosecution. Lastly, it may be noted that in the TSR team's response report it was observed, at para 3.24, that there was a possibility that the setting up of a separate investigation body might lead to a loss of expertise to the HMRI and even more fierce competition for recruits.

Conclusions

9.29   The criticisms of the HMRI which I have outlined in para 9.26 relate essentially to matters of performance which are capable of being corrected at some cost and given suitable direction. There is, on the other hand, a strong argument for an investigating body which enjoys real and perceived independence. The recognition of the strength of those arguments is demonstrated by well established arrangements in regard to aviation and transportation systems in other countries. There is also force in the added benefit which concentration on the learning and application of the lessons for accidents and incidents can bring. Against this I have to weigh the potential disadvantages, the most significant of which is the loss of direct connection between the investigator and the regular contact with the operation of safety systems. However, on balance I consider that the stronger arguments are in favour of change, and I accordingly recommend that the responsibility for the investigation of accidents should be entrusted to an independent body which is set up for the purpose. The body would be similar in constitution to the AAIB and the MAIB. For convenience I will refer to it as the RAIB. I will discuss this in more detail in Chapter 11. These remarks apply to investigations currently carried out under Section 14(2)(a) of the 1974 Act. As regards Section 14(2)(b), the work of the RAIB may reduce the need for inquiries of the type for which it currently makes provision. However, in future there may be accidents or other matters of such a nature as to call for them. Given my recommendations in regard to investigations corresponding to Section 14(2)(a), it would be appropriate that the appointing body for inquiries corresponding to Section 14(2)(b) is the Secretary of State.

9.30   I have not so far referred to the implications of a trans-modal body for the investigation of accidents and incidents. This subject lay beyond the province of this Inquiry and none of the parties took the opportunity to advocate that it should be pursued. However, the Joint Rail Unions indicated that they favoured it as a long-term objective. They pointed out a number of possible advantages, including the benefit of synergy between the investigation of different modes of transport, the opportunities for carrying out common studies into subjects such as fatigue, the encouragement of the extension of anonymous reporting, the development of specialism, greater emphasis on wider lessons, and a higher profile for the checking of progress on the implementation of recommendations. A number of examples of trans-modal bodies were mentioned in the evidence, including the USA's NTSB. In the circumstances I make no recommendation in regard to advancing in the direction of a trans-modal body. However, as Counsel to the Inquiry pointed out, the establishment of the independent investigatory body which I have recommended in this chapter will not conflict with a cross-modal body if in due course the Government decided that the latter should be pursued.

A new safety regulator?

The models

9.31   ATOC, the ROSCOs, the SRA and Railtrack, along with the bereaved and injured represented by the Southall and Ladbroke Grove Solicitors' Group and the Rail Users' Committees, advocated the creation of a new safety regulator for the railways in place of the HSE (see Appendix 5). Each of these proposals, apart from that of Railtrack, entailed that Railtrack should cease to exercise a role which was described as "quasi-regulatory", a proposition which I have already accepted in Chapter 8, and accordingly to that extent I do not require to cover that point again in this chapter.

9.32   One of the leading exponents of the proposed new safety regulator was ATOC. They submitted that this proposal would:

(i)   provide a clear delineation between the regulator and the regulated, along with a clear indication that the setting of Railway Group Standards and their enforcement were regulatory functions;

(ii)   enable actions and recommendations to be collated and prioritised;

(iii)   simplify and streamline the processes for approvals and external auditing;

(iv)   assist in bringing about beneficial changes in safety culture; and

(v)   assist in giving strategic safety leadership.

9.33   ATOC submitted that, as at present, the responsibility for the safety of Railtrack's own operations would remain with their Assurance and Safety Directorate. The rights and duties of parties *inter se* would continue to be governed by the track access agreements between them. The product of the monitoring and audits of the NRSA, the new safety regulator which they advocated, would be made available to all relevant industry stakeholders. It was envisaged that there would be an on-going cycle of feedback.

9.34   It was emphasised by ATOC that their model was consistent with the principles which had been endorsed by the Robens Committee. Primary responsibility for safe operations would remain with those conducting and controlling them. Regulations and standards could contain a range of provisions from the goal-setting at one end of the spectrum to the highly prescriptive at the other.

9.35   ATOC maintained that their model was clearly workable. The evidence did not suggest otherwise. Similarities with the regimes in Canada, New Zealand and some of the states in Australia were pointed out. It was consistent with what was indicated as the lines on which a future European Directive on safety might be based. It had worked well in the case of the CAA (see Appendix 6). It would ensure a genuinely fresh start and enhance prospects of recruitment. It would operate in much the same way as the CAA. It would, however, be essential that the NRSA were

"…strong, well-resourced, proactive, and knowledgeable, impartial and able to provide strategic direction aligned with SRA and industry objectives".

9.36   Counsel for the bereaved and injured represented by the Southall and Ladbroke Grove Solicitors' Group submitted that the body with the best knowledge of the failings of duty holders, which it obtained through inspection and audit, was best placed to impose sanctions. The CAA model was very attractive in its requirement for comprehensive certification. ATOC's actions in drawing up Codes of Practice for train driving demonstrated the need for co-operation and standardisation in order to mitigate the dangers inherent in privatisation. All industry parties should report to the NRSA any defects which they had found in the course of monitoring. As regards the imposition of penalties, and the offering of incentives, with regard to safety performance, it might be best if the Rail Regulator were required to put his powers at the disposal of the NRSA when they so requested.

9.37   The SRA said that a safety regulatory body on the model of the CAA was worthy of serious consideration. The railways required a fresh start in the process of safety regulation.

9.38   The models proposed by these parties raise two important issues, namely:

(i)   the responsibility for Group Standards; and

(ii)   the choice of safety regulator for the railways.

I will consider each of these in turn.

9.39   Each of the models for a new safety regulator, apart from that proposed by Railtrack, involved that the new safety regulator would assume responsibility for Group Standards. This was clearly regarded by proponents as appropriate territory for a single regulatory body. They drew a parallel with the regulatory control exercised by the CAA.

9.40   In the treatment of industry standards there are two possible models. In one model – of which the CAA are an example – the safety regulator takes charge of industry standards (and likewise the licensing of equipment, premises and personnel). In the other an industry board is in charge of such standards, subject to the scrutiny of the safety regulator, and the latter is responsible for regulations and high-level regulatory standards where this is required in the public interest and the industry is considered to be in need of guidance or direction. It should, of course, be noted that neither of these models corresponds to the present situation in which industry standards are treated, at least by Railtrack, as the means by which they discharge their duty to control the importation of risk onto the network. What is appropriate for one transportation regime may not be appropriate for another, and accordingly I require to look closely at the characteristics of the rail industry in Great Britain as it is presently constituted.

9.41   At the outset it is important for me to bear in mind the origin and nature of Railway Group Standards. They are the successors to the standards which were evolved by British Rail for their own internal purposes. Although they have been considerably improved as a result of the work done by the S&SD, they remain in essence standards which are internal to the industry itself. As Counsel to the Inquiry pointed out, under reference to the evidence of the Rail Regulator and Mr Beswick, the standards relate not only to matters of safety but also to procedures, processes and other matters to do with harmonisation which have an economic objective. Accordingly, it is not straightforward to identify those standards, or those parts of standards, which would be appropriate for a safety regulator to make a part of its responsibility.

9.42   Each of the models which was proposed plainly recognises that in the process of arriving at a Group Standard which they endorse, the NRSA would be dependent in some degree on the advice which is given by the industry. Thus the ATOC model included a "Rail Standards Setting Executive", which would be a rationalisation of the existing RISSC and the standards subject committees, and would draft standards for submission to the NRSA. In the model put forward by the Southall and Ladbroke Grove Solicitors' Group the standards to be sanctioned by the NRSA would be drafted by a pan-industry body, while the NRSA themselves would have the power to amend or to draft and impose standards. In the case of the SRA, Mr Morris, Executive Director of London South East and formerly the Technical Director of Safety and Operations for the SRA, accepted that there would have to be "a huge element of contact" between the safety regulator and the operators. "So I would see", he said,

> "...a lot of liaison between these two particular parties in the setting of those Group Standards but the final arbiter would be the regulatory safety body".

I am in no doubt this association between the safety regulator and the industry would be a matter of practical necessity.

9.43   It is, however, necessary to consider the full implications of such an arrangement. First, it clearly recognises that the knowledge, experience and expertise for the formulation of industry standards reside in the professionals who are involved daily in the work of the industry. Could a body which was not an industry body match that knowledge, experience and expertise? Secondly, in any event could it do so without wasteful duplication of the work of the industry? Whether either of these questions can be answered in the affirmative is doubtful. I regard the suggestion that the NRSA would themselves amend or draft standards as being unrealistic.

9.44   The models which have been proposed envisage that, on the one hand, the safety regulator would be the body which had responsibility for sanctioning standards; and, on the other hand, it would maintain a considerable amount of contact with the industry. It seems to me that this carries with it the danger of too close a relationship with the industry. Counsel for the Joint Rail Unions pointed out, with some justification, that this did not sit well with ATOC's assertion that there should be a clear delineation between the regulator and the regulated. Mr Rimington observed that such an arrangement

>   "...would threaten to involve the regulator too closely in the affairs of the industry and so in the end threaten its independence. Railway industry standards are very closely bound up with operational practicalities and only to a varying extent with safety; they are in a very real sense the business of the industry itself".

I am in no doubt that if a safety regulator is to discharge its own distinctive role properly, it has to be distanced, and be seen to be distanced, from the industry and its members. Thus, for example, if the safety regulator had sanctioned standards, would it be able, or be seen to be able, to exercise adequate scrutiny of the standards or the processes for setting them? In regard to an appeal by a railway operator would it be, or be regarded as, free from bias? I also agree with the submission made by Counsel for the Rail Regulator when he said, referring to the assumption by the safety regulator of responsibility for Railway Group Standards, that

>   "...combining the two authorities deprives the process of a necessary and very valuable interaction or even tension which would be the spur to constant renewal and improvement of the standards".

Counsel for the HSE drew attention to a further matter which is of some concern to me. He pointed out that if the NRSA were staffed by industry personnel, whether on secondment or otherwise, there was a risk of "regulatory capture". This was defined by Mr V P Coleman, Chief Inspector of Railways for the HMRI, as a situation in which the regulator becomes so close to, or closely involved or associated with, what is being regulated that there is little practical difference between the stance of the regulator and that of the regulated.

9.45   If the standards are to be set by a single authority, there is also, in my view, some risk that they would become, as the Rail Regulator submitted, more closely akin to detailed regulations, and hence more likely to be prescriptive rather than goal-setting in nature. Counsel for the HSE emphasised that prescriptive standards were less demanding and less flexible since they presented only minimum standards. There was, he said, a danger of a "light touch", which the HSE now regarded as inappropriate.

9.46 These considerations, taken in combination, convince me that it is inappropriate that a safety regulator for the railways should assume responsibility for Group Standards.

The choice of safety regulator

9.47 I now consider the central feature for each of the models with which I am concerned in this chapter – the proposed replacement of the HSE by a new body as the safety regulator for the railways. As can be seen from Appendix 5 there are significant differences between the characteristics of the various models which were proposed. In particular the model proposed by Railtrack differs markedly from that of the SRA, in respect that it did not incorporate functions relating to Group Standards, despite the fact that Railtrack, who came as late converts to the idea of a new safety regulator, claimed that their attitude had been influenced by the evidence of Sir Alastair Morton. Chairman of the SRA. In passing I note that the Railtrack model is said to have arisen from a discussion between the Chairman and the recently appointed Chief Executive, apparently without involving Mr R I Muttram, Director of the S&SD.

9.48 The parties who advocated a new safety regulator put forward a number of arguments.

9.49 First, they relied on criticisms of the past performance of the HMRI. In my report on Part 1 of this Inquiry I stated that the criticisms expressed in the Internal Inquiry Report were well founded (para 10.22). These were directed to the time taken for the approval of the resignalling scheme, the slow progress in bringing issues to a conclusion, and inadequate risk analysis. Miss Bacon candidly accepted that more could and should have been done to enforce health and safety legislation. In this part of the Inquiry I was referred to the HSE's General Issues Report which considered the HMRI's general approach to regulating the rail industry. This report found weaknesses in the internal systems of the HMRI which "contributed to a rather slow response to challenges posed by the changing industry". This led, it was stated, to failures to make timely responses, less than appropriate action and inadequate communications between the different parts of the HMRI. The conclusions of this report in regard to the second of these matters is of some significance, in particular its comments on the HMRI's use of a "light touch" approach to enforcement and action taken. The report stated:

> "HMRI has followed a consensus seeking approach to dealing with duty holders, with a reluctance to use formal enforcement powers. The 'light touch' approach was derived during a political climate of deregulation and at a time when the true nature of the reorganised industry was still emerging. Such an approach is less appropriate for an industry which is more driven by commercial imperatives and in which levels of co-operation and trust appear to be declining".

I also note that the report by DuPont Safety Resources on Safety Management in the Railway Group stated that it had found that in some areas the HMRI had been too distant till a crisis occurred. However, a notable exception was the HMRI's report on

SPAD management in September 1999, to which I referred in para 10.16 of my report on Part 1.

9.50   ATOC also criticised the HMRI for reluctance to commit themselves where they should have given a clear lead. They had given a half-hearted response when the Rail Regulator consulted them in September 2000 about his proposal to double the penalties for trains being delayed. They had not committed themselves to a level of expenditure to represent a value per fatality. ATOC also criticised the HMRI for lack of consistency in regard to the "touch" which they applied, and for confusing their avowed adherence to the ALARP principle.

9.51   Secondly, Sir Alastair questioned

"…whether the cause for regulation has been well served by the evolution of rail safety regulation since the mid-1990s. I refer to the diffusion of railway regulatory activity and support across the HSE as a whole, coupled with a worrying decline in in-house rail experience. There is also the question whether, in the process, the HSE has become defensive".

He said that such problems are

"…not likely to be resolved within the existing organisation because it has all the baggage of an existing organisation".

Among his comments about the way safety is managed on Britain's railways he described the current climate as

"…sour and defensive, not least between regulator and regulated, but also between employer and contractor or purchaser and supplier".

Both the SRA and Railtrack emphasised that the attraction of an adequate number of personnel of high calibre was hampered by the safety regulator being within a Governmental body which had wider responsibilities.

9.52   Thirdly, Sir Alastair advocated a new institution

"…because I believe the task is highly specialised and, in today's system, very complex. Britain's rail network is very dense, carries more high speed trains than most, if not all, others and now involves more and more technology akin to aviation, away from rail's steam heritage".

This led, he said, to the SRA urging the Inquiry to "read across" from the example of the CAA. He questioned why aviation should have "a much more intellectual approach to safety and management than railways". He therefore favoured a fresh start in the process of safety regulation. I might add that he clearly believed that a new safety regulatory body should be responsible for the setting of standards, adding that this entailed that the latter had the respect of the industry, stating:

"…this new regulatory body and the industry must be looking each other in the eye with confidence in each other".

9.53   There is no doubt that there have been significant deficiencies in the HMRI's performance, as described in their own Internal Inquiry Report and General Issues Report. It is clear that one of the factors has been the shortage of resources. However, there are other deeper causes. There has been a failure to adapt to the changing conditions brought about by the disaggregation of the rail industry and the barriers created by competing commercial interests. The evidence also gave me a strong impression of a difficult relationship with Railtrack, largely due, in my view, to the latter's dominant position. Railtrack originally described themselves in their safety case as being "the directing mind" of the railways. They later departed from this description, with the consent of the HMRI, but asserted their full control over the conditions of use of their network, while at the same time declining to give the safety leadership to the industry and disclaiming that they were responsible for ensuring safety on the network or had power to do so.

9.54   I have had some difficulty in relating the remarks by Sir Alastair about rail's steam heritage to the position of a safety regulator. His remarks did not appear to recognise that a safety regulator is not responsible for the management of safety. This is no academic distinction since the effectiveness of the work of a safety regulator depends on its ability to subject operators' management of safety to independent scrutiny. Sir Alastair's imagery was arresting but I am not impressed with its relevance to the question which I have to consider.

9.55   As regards the expertise of the HMRI, I am in no doubt that their ranks include specialists who are entirely at home with the technological advances in Britain's railways. More generally, the HSE are, I accept, no strangers to the regulation of specialised, complex and technologically advanced industries in which there is the potential for major hazards. However, I have concern that the recent "brigading" of HMRI has not addressed the need to build up the numbers of personnel who have relevant experience in regard to the rail industry itself.

9.56   In considering the arguments in favour of a free-standing safety regulator for the railways, I have to set these against the arguments in favour of the HMRI as part of the HSE. They include the cross-fertilisation of ideas, the sharing of technical resources and the support of a well-developed regulatory framework. The Joint Rail Unions drew attention to the importance which the Robens Report placed on the learning that takes place across industries between complex sectors, which was said to outweigh the benefits of single issue, single industry bodies. Mr Coleman, describing the risk of "regulatory capture", emphasised that

        "...being part of a larger organisation which is not centred wholly, squarely or even in a majority way on a particular industry and where we get transfer of staff in and out on dealing with particular industrial sectors, makes the propensity rather less".

It is clear that the HSE have a well-established and well-recognised independent stance. I noted that Dr R A Cox, Consultant Engineer, said that his reason for saying that the railway safety regulator should stay within the HSE was that

"...the railway industry generally has suffered from a problem of insularity in the field of safety culture and there is a need for cross-fertilisation between that industry and the best of practice in other industries".

He said that because of their cross-sectoral scope the HSE were the only body which offered that opportunity, which he felt had not been fully exercised to date.

9.57 It is wise, in my view, to be cautious about drawing parallels between the CAA and the safety regulator for Britain's railways. As was pointed out by Sir David Davies, President of the Royal Academy of Engineering, in the case of railways interactions between trains and between them and the track are of critical importance. They do not have a parallel in aviation. Also I have noted that the regime of the CAA requires to fit into a world-wide system for the control of aviation. At an operational level there is a considerable amount of prescription in the regulatory controls which are imposed. Mr Muttram also pointed out:

"Airworthiness is almost uniquely contained within the vehicle. And whilst the FAA and the CAA set the high level standards very similar to the HSE's principles and guidance, most of the detailed technical standards for aircraft are actually written by the aircraft manufacturers. It is a completely different situation".

The scope of licensing by the CAA is extensive. At present there is little corresponding to this in the safety regulation of railways. Mr Rimington described the retention by the HMRI of direct responsibility for the approving of railway equipment as "somewhat anomalous". The HSE emphasised that this did not involve the HMRI in verification: it was a "confirmation of due process", i.e. that the equipment was produced according to an accepted methodology. Miss Bacon indicated the view that it would be better for others to be responsible for approvals under the aegis of the safety regulator, rather than that the latter should be directly involved. At present this matter was out for consultation, invited in the HSE's discussion document "Regulating Higher Hazards: Exploring the Issues". Mr Profit very fairly accepted that while the system of his organisation had great clarity, he was not sure that it would be an appropriate model for the rail industry.

9.58 It was plain from the evidence of Mr Morris, who spoke to the case for the SRA, that he tended to favour the prescriptive approach. He accepted that this represented a "fundamental difference" from the approach of the HSE and the Robens Committee. It is well recognised that a degree of prescription may be inevitable as a matter of practical necessity or as a matter of policy. However, an undue extension of prescription carries with it the risk that duty holders will not be alert to the need for regular appraisal of risks and the adequacy of the measures for controlling them.

9.59 The Railtrack version of the proposal for a new safety regulator (see Appendix 5) rightly attracted criticism from Counsel for the HSE for displaying vulnerability to "regulatory capture". Mr Coleman said that it was much more likely to happen

"...if you have a group of people who share a common basis, that have all grown up the same way in the same industry and are involved very narrowly in particular regulatory areas".

Mr Rimington observed:

> "All industries desire to have their 'own' safety regulator, all desire to 'capture' him, historically some have done so; and the smaller and weaker the body is, and the more isolated from other influences, the more likely it is to happen".

9.60   The General Issues Report indicated that the HSE believed the days of the "light touch" were over. It is recognised that the HMRI are still under-resourced, despite considerable recent growth. Mr Coleman said in evidence that the HSE were seeking to double their size by 2002. In the course of her evidence Miss Bacon said that

> "...we are looking at a situation where clearly there is deep discontent, understandable discontent, about the level to which duties are being complied with and perhaps concern about whether we can ever reach the point where those duties are going to be complied with without a greater regulatory intervention. I will be more comfortable with a greater degree of intervention than we have had over the past five or so years since privatisation, because I think there are some problems to be resolved, and at least until we can be more confident about the competence levels available in the industry, the way that resources are deployed and safety culture within the industry, then it is right to be going for a more interventionist and intrusive regulatory regime with all the resource cost that that implies. I think it can be done within the Health and Safety at Work Act framework".

She emphasised that the Act was extremely flexible, as it could accommodate a variety of regulatory techniques. The choice as to how the regime should develop was

> "...fundamentally a political decision in the end about the intensity of regulation, the intrusiveness of the regulation".

Conclusion

9.61   I can readily see that there are presentational attractions in the institution of a new single safety regulator with functions covering the range illustrated in the models, such as that advocated by ATOC. However, there seem to me to be significant drawbacks, in particular the difficulty of the safety regulator maintaining, and being seen to maintain, its independence. Public confidence in its independence is extremely important. The function relating to Group Standards is but one example of aspects to which this applies.

9.62   I am not convinced that the CAA provide a helpful parallel for railways.

9.63   Two features of the models call for additional comment. First, a number of them envisage that the new safety regulator has the functions of establishing and managing system authorities and of funding and sponsoring research and development. These functions do not strike me as providing an appropriate fit with the role of the safety regulator. They rather belong to the industry itself.

9.64   Secondly, I note that, while most of the models, apart, of course, from that of Railtrack, envisage that Railway Safety would be superseded, they hardly address certain existing functions of Railway Safety which are of significant importance, namely:

(i)   the monitoring and reporting of the industry's safety performance; and

(ii)   the development of the annual Railway Group Safety Plan which complements Railtrack's Safety Plan (see para 3.11).

As regards the latter, Mr C Carr, Technical Director of Amey Rail, said:

"I think it is a very important process that does encourage the different parts of the industry to recognise that there are common objectives to improve safety within the railway and that we must all pull from our different directions to move those objectives forward".

Mr Taig commented:

"I feel it represents the industry taking ownership of issues and proposing goals for itself".

While, as Mr Taig pointed out, the safety regulator would be consulted during the development of the Plan, it is plain that it must be free to say whether the industry is aiming too low or not aiming at the right target. Mr B R Burdsall, Managing Director of Midland Main Line, hazarded the view that the NRSA would handle the Plan, but it is plain, in my view, that the function of developing the Plan is not one for the safety regulator since it would compromise its need to express an independent position. In these circumstances it is clear that the function relating to the Plan and to the monitoring and reporting of safety performance of the industry would still require to be administered by some separate body and accordingly there is not the degree of simplicity in the ATOC model that might first appear.

9.65   The deficiencies in the past performance of the HMRI, while significant, do not seem to me to reveal a question of principle as to whether the HSE are an appropriate body to be the safety regulator. Nor do they show that the HMRI are not capable, given adequate resources and effective leadership, of adequately discharging that function as part of the HSE. They have shown a capacity for self-appraisal, and are applying the lessons of past failures. They are reviewing the extent to which they should recommend that a more intrusive and interventionist approach should be taken and, in the light of that, the extent to which they require additional resources.

9.66   Having regard to the foregoing, I am not persuaded that a new safety regulator for the railways should take the place of the HSE. I recommend that the HSE through the HMRI continue to fulfil that function. However, I would emphasise that it is imperative that the HSE are provided with adequate resources in order to fulfil their role, whether or not that role takes the more intrusive and interventionist form as was envisaged by Miss Bacon. The recommendations which I make in regard to strengthening the role of the safety regulator in regard to the assessment of safety cases and their revisions (para 8.36) and in regard to auditing (para 8.37) plainly have resource implications. Further, it is extremely important that the "brigading" of the HMRI should not distract attention from the need of the HMRI to recruit and maintain personnel who have the relevant experience and expertise in regard to the railways. A number of parties rightly drew attention to the need for specialist training and graduate recruitment. Lastly, but by no means least, while I welcome the formation of a railways directorate within the HSE, it is plain that there is a need for the existing leadership of the HMRI to be reinforced through the addition of new imagination and energy. I recommend that it should be placed under the direction of a new post, to be filled by a person of outstanding managerial ability, not necessarily with a railway background. This post should be regarded as commanding a special salary level for the purpose.

European regulation.

9.67   I also require to consider whether my conclusion would be consistent with European regulation.

9.68   During the evidence of Mr Muttram the attention of the Inquiry was drawn to the Common Position No 41/2000 adopted by the Council of Ministers on 28 March 2000 in regard to the development of the railways of the European Community. According to its terms it was proposed that Article 7(2) of Council Directive 91/440/EEC should be replaced with the following:

> "Member States shall ensure that safety standards and rules are laid down, rolling stock and railway undertakings are certified accordingly and accidents investigated. These tasks shall be accomplished by bodies or undertakings that do not provide rail transport services themselves and are independent of bodies or undertakings that do so, in such a way as to guarantee equitable and non-discriminatory access to infrastructure".

I note that these terms would not require that all of these functions should be discharged by the same body. In particular they do not rule out the setting of standards, such as Railway Group Standards, by a body other than a national safety regulator for the railways, so long as that body did not provide, and was independent of the providers of, transport services, i.e. in the British context, TOCs.

9.69   It is also important to note that NERA in their report in February 2000 to DG Energy and Transport, to which I referred in para 9.25, stated in para 5.3.2.2:

> "Much use is already made in Member States of international standards. However the usual railway practice is for standards and decision rules to be interpreted and imposed by the Railway Inspectorate (RI) and/or the infrastructure manager. This is an obstacle to innovation. In other industries, the responsibility for fulfilling the general legal obligation normally rests with the industry. The safety regulator's role is to supervise this, not to do it. We believe this should be the case with the railways.

> This implies a national body which represents the infrastructure and train operators and the supply/maintenance industry. This body, say a Railway Industry Safety Committee, with a supporting executive staff, would also ensure that adequate R&D on system-wide safety issues was undertaken and promote a

forward looking approach, for example on national policy towards European harmonisation.

The standards and decision rules proposed by the industry need to be subject to veto by the RI. Trade union and consumer representative input should be through the RI, not through the Railway Industry Safety Committee".

In his written statement Mr M Spackman, the leader of NERA's project team, included passages from the NERA report, such as the passage which I have quoted above. His statement was incorporated into the evidence before the Inquiry on 7 December 2000.

9.70   The letter from Mr Hilbrecht, dated 20 October 2000, to which I referred in para 9.25, stated that the planned European Directive on Railway Safety was intended to deal with, *inter alia*, "the regulation of the appointment of independent national bodies in the Member States responsible for regulating, monitoring and enforcing the overall safety performance of the railway system. Provisions will be made for co-operation by national bodies under the supervision of the Commission". For the HSE to remain the national safety regulator for the railways in Britain appears to be consistent with this. While the letter made no explicit reference to standards such as Railway Group Standards, for them to be set by an industry standards body in the way described by NERA in their report to DG Energy and Transport did not appear to conflict with what was described in the letter.

9.71   In July 2001, at a stage when the writing of this part of my report was well advanced, I was informed that the European Commission had produced a draft Working Document dated 22 June 2001 for the proposed Directive. I understand that this is likely to lead to a formal consultation draft Directive later in 2001. Article 12(1) of the document proposes a national safety authority, as follows:

"Each Member States *(sic)* shall establish a safety regulatory and supervisory body, a national safety authority. It shall be independent in its organisation, legal structure and decision making from any railway undertaking, infrastructure manager, charging body, allocation body or applicant"

Article 12(2) proposes that this safety authority should be entrusted with a number of regulatory tasks, including

"…the issue of safety rules under national law as described in Article 5(3) of this Directive, unless the rules are issued by the national Parliament or the Government".

I note that Article 5(3) refers to a number of categories of rules, including

"…the general operating rules of the railway network that are not yet covered by a TSI (Technical Specification for Interoperability), including rules relating to the signalling and traffic management system",

cf Article 5(2).

9.72 So far as I am able to determine, for the HSE to remain the safety regulator for the railways in Britain would be consistent with the document. However, the implications of the proposals about safety rules such as "the general operating rules of the railway network" are not clear. The proposals appear not to have been framed with specific reference to the position in Britain. It is not clear whether they are based on any comprehension of the nature, range and application of Railway Group Standards in regard to safety or non-safety matters, and it may be that they are intended to address matters in other countries which have no counterpart in Britain.

9.73 I have considered whether the parties to the Inquiry should be invited to give their views on the Working Document, but I have decided that this is not necessary; their respective positions were clearly set out in the course of the hearings of Part 2. This Inquiry proceeded on the basis that it was for it to consider, and make recommendations as to, the future of the functions which were exercised by the S&SD, such as those relating to Group Standards. As I noted in para 8.12, the Deputy Prime Minister stated on 23 February 2000 that the "final fate" of the S&SD would be decided following the report of this Inquiry. I am concerned to ensure that, so far as I can, any recommendation which I make, such as in regard to the future of the standard setting function previously exercised by the S&SD, is capable of being put into effect. Its future was, of course, one of the main issues in the Inquiry.

9.74 In the light of the above and my conclusions from the evidence, I recommend to the Government they should use all reasonable endeavours to ensure that standards such as Railway Group Standards are not required by the Directive in its final form to be set by the safety regulator, and that the draft Directive is modified to such extent as is necessary for that purpose.

The safety regulator and the public

9.75 The public perception of safety in a transportation system such as the railways can have a significant influence on what is expected of the safety regulator. The Inquiry had the benefit of the views contributed by the participants in a seminar on Public Perceptions of Rail Safety, and the evidence of Mrs Deirdre Hutton, Vice Chair of the National Consumer Council.

9.76 One of the main points made at the seminar and by Mrs Hutton was that the public have difficulty in understanding the structure of the rail industry and its regulation. I have no doubt that the public are entitled to expect that it is clear who is responsible for what. To an extent the complexity is an inevitable consequence of privatisation and disaggregation of the industry. It leads, for example, to the need for economic regulation and the regulation of franchises and public funding. However, some improvement in clarity should be achieved by the implementation of my recommendations in regard to the ending of "quasi-regulation" by Railtrack, the responsibility for Railway Group Standards and maintaining the independence of the safety regulator from the rail industry and its constituent companies.

9.77 Another important aspect that was highlighted at the seminar and in Mrs Hutton's evidence was the public's confidence in decisions. Mrs Hutton emphasised that it was important that the public should know that a body which had "clout" took their concerns into account when decisions were being made.

9.78   At present the HSC are advised by the RIAC, as I noted at para 3.68. The Inquiry was informed by Mr Coleman, who has chaired the RIAC since 1998, that they comprised representatives of Railtrack, LUL, the three main railway trade unions, passenger groups (being the Rail Passengers' Council and the London Transport Users' Committee), ATOC, rail freight interests, the RIA, ROSCOs, the Confederation of Passenger Transport (representing light rail and tramway interests) and the Heritage Railway Association. The HSE and the DETR (now the DTLR) act as observer and assessor respectively. He said that it was also intended to add representation from Railway Safety. He described the RIAC as

"…the body with the most widely representative membership of any formal body dealing with safety issues in Britain".

They were set up to advise the HSC and provided

"…a very important forum for channelling the views, collectively, of all these various stakeholders through the Commission".

He added that clearly the HSE also benefited from that process. The minutes of the RIAC were put onto the internet. The RIAC discussed key issues, but could probably do more. He also pointed out that it was recently decided that they would meet in public, as a means of improving public understanding of their work.

9.79   Dr Cox spoke in favour of the setting up of a Rail Safety Advisory Commission which was advocated by the Collins Passengers' Group (see para 36 of Appendix 5). Dr Cox envisaged a body comparable with the Human Genetics Commission or the Biotechnology Commission, which were new bodies "with a great deal of clout". He envisaged that it would be established under statute. It should exercise a supervisory role in relation to the HMRI, meaning by this that the HMRI would have to give weight to its views. It would report to the HSC (and possibly also the SRA). It would be powerfully representative of "those who bear the risk", but it would also be representative of the rail industry. Dr Cox regarded the RIAC as not having the same balance. Further, their agenda was more focused on tactical rather than strategic issues. He accepted that what he had in mind was equivalent to a strengthened and elevated RIAC.

9.80   While I see force in the view that steps should be taken to strengthen the communication of the passengers' viewpoint and to ensure that it is expressed on matters of rail safety at a high level, I am not convinced that it is necessary to establish an additional statutory body for the purpose. There have been concerns that the Rail Passengers' Council and their affiliated committees have shown less activity in regard to safety than performance. Whether that is so or not, I consider that the extent of passenger representation on the RIAC should be re-considered, with a view to increasing the number of those who represent the interests of passengers, but not necessarily drawn from those bodies. It is, of course, important to bear in mind that this Inquiry was assured by Mr Coleman that all members of the RIAC had the opportunity to contribute fully. The RIAC did not proceed by voting. I also draw attention to the remarks of Dr Cox in regard to the level of the involvement of the RIAC. I support the view that they should be concerned with questions of safety strategy at a high level.

The relationship between the safety regulator and the Rail Regulator

9.81   While the Inquiry tabled for discussion the question of whether the safety regulator and the economic regulator should be parts of a single body, none of the parties to the Inquiry sought to argue for this. It was, however, put forward as part of a submission made to the Inquiry on behalf of the Conservative Party. I am satisfied that the respective functions of the safety regulator and the Rail Regulator, which are clearly distinct, should be seen to be performed by different bodies. This is desirable in order to give public confidence of their independence and of their distinct accountability. Neither is there a good case for a new authority above both of these regulators and the SRA. The functions are fundamentally different. Moreover, while the SRA are subject to Government direction, the Rail Regulator is not.

9.82   As I noted in para 3.43 in a brief outline of the functions of the Rail Regulator, he has a duty under Section 4(3)(a) of the 1993 Act, in exercising his functions,

> "…to take into account the need to protect all persons from dangers arising from the operation of railways, taking into account, in particular, any advice given to him in that behalf by the Health and Safety Executive".

The evidence highlighted the desirability for closer alignment between these two regulators.

9.83   The Rail Regulator said in evidence that a review of the relationship between his office and the HSE had led to an increase in the team supporting Mr Beswick, the Director of Network Regulation, in regard to safety. There had been improvements in the nature and quality of the arrangements for liaison with the HSE, and in the exchange of information between the two organisations. He had worked very closely with the HSE in regard to such matters as Railtrack's stewardship of the rail network and their plans for dealing with broken rails. There was sufficient and satisfactory co-ordination between his office and the HSE. He was, on the other hand, dismissive of the suggestion by Railtrack that there should be increased transparency in the details of his liaison arrangements.

9.84   In his statement of case the Rail Regulator stated that he required to be satisfied, before taking any decision which had an impact on safety, that he had

> "…a robust and realistic safety assessment from an independent body, so that whatever decision he makes will not adversely affect safety".

He also said that he must have confidence in the advice which he received. I appreciate the practice which the Rail Regulator has followed, which is consistent with the terms of Section 4(3)(a) of the 1993 Act. A duty to take into account the advice of the HSE is, of course, only an aspect of a general duty

"…to take into account the need to protect all persons from dangers arising from the operation of railways".

From what I have set out in this report it will be obvious that the powers of the Rail Regulator can be exercised to substantial effect in the area of performance. In my view, there is considerable force in the submission of Counsel for the bereaved and injured represented by the Southall and Ladbroke Grove Solicitors' Group that proposals by the Rail Regulator which have a potential to affect safety should be the subject of a risk assessment. It should be the responsibility of the Rail Regulator, as the proposer, to show that his proposals would not adversely affect safety. Whether he obtains such an assessment from the HSE or from some other competent body is a matter for him.

Appeals in respect of Railway Group Standards

9.85   As I noted in para 3.51, an appeal in regard to the setting of a Group Standard lies ultimately to the Rail Regulator. In his Notice of Modification to Condition 3 of Railtrack's network licence the Rail Regulator has indicated that he considers it appropriate, pending the outcome of the present Inquiry, that he as economic regulator should remain the appeal body. He pointed out that the licence condition made it clear that he would consult the HSE in the event of an appeal.

9.86   In the circumstances I see no good reason why the Rail Regulator should not remain responsible for appeals, on the basis that, as at present, he seeks the advice of the safety regulator on safety matters.

The enforcement of Railway Group Standards

9.87   The enforcement of Group Standards presents a more complicated situation. The Rowlands Report at para 42 pointed out, in the context of a safety-related Group Standard, that there was a confusing overlap between the jurisdiction of the Rail Regulator and that of the HSE. The former, I note, can enforce such a standard under the 1993 Act applying one set of possible sanctions, while the latter can enforce it under the 1974 Act with a different set of sanctions.

9.88   In practice this overlap has been dealt with under the memorandum of understanding between the Rail Regulator and the HSE. In their evidence the Rail Regulator and Mr Beswick explained that the HSE would normally be the body taking action in respect of the safety element of Group Standards. If the HSE asked the Rail Regulator to take action because his powers were more suitable, he would expect to do so. On the other hand, to the extent that a Group Standard contained matters of a purely economic nature, the Rail Regulator would expect to take responsibility. As Mr Beswick pointed out, Group Standards include

"…purposes which are essentially economic purposes about efficiency, economy, benefits to users, that sort of thing".

So far there has been only one instance in which the Rail Regulator took action and that was in respect of a non-safety matter. This arose out of a complaint by manufacturers against Railtrack in regard to the supply of information about the gauge and condition of the network.

9.89   The Rail Regulator made it clear that it was his view that he should not be enforcing safety standards. He also said:

> "It is inappropriate for me to set up a parallel safety enforcement staff. Indeed the statute requires me to take advice on safety matters from the HSE. So the memorandum of understanding provides that the HSE is the eyes and ears of the Rail Regulator in matters of safety. That is how it works at the moment. It is not perfect, but it does work".

9.90   In his consultation document in regard to the modification of Condition 3 of Railtrack's network licence the Rail Regulator put forward a proposal that all licences which required compliance with Group Standards should be modified to require the licence holder to comply with such Group Standards or parts of such standards as might be specified in a notice issued by him to the licence holder, after consulting the licence holder and the HSE. In his Notice of Modification he recorded that a number of consultees had stated that there was insufficient clarity in what was proposed. In the circumstances he concluded that he should not pursue this proposal for the time being.

9.91   While I can well understand the concern expressed by the Rail Regulator that a way should be found by which he should be relieved of the duty to enforce what are in substance safety standards, it is not easy to devise a clear and effective method for the purpose, since disentangling non-safety-related Group Standards or parts of such standards is not straightforward. For the reason given in para 2.27 of this report I make no recommendation as to the way forward, but suggest that this question, which does not seem to be of a pressing nature, should be the subject of further study.

The HSE's advice in regard to the periodic review

9.92   The Rail Regulator said that he gave "very great weight" to the advice given by the HSE. In the case of the periodic review, if the HSE had ever raised material concerns in relation to any aspect of it, he would have given them very considerable weight.

9.93   This is an area where the evidence gives me some concern. The attention of the Inquiry was drawn to a letter from Mr Coleman to Mr Beswick dated 17 October 2000 with reference to the Rail Regulator's proposals for the periodic review. In the course of that letter Mr Coleman stated:

> "The proposed improvements to efficiency should not lead to any pressure or tendency to affect the control of risk. We have discussed and agreed in the past that safety must not be compromised in any quest to improve performance and efficiency".

The Rail Regulator took this letter as a "green light" in relation to the periodic review, including proposals for the doubling of incentives and the abolition of the free possessions allowance. Mr Coleman did not dispute the HSE had not objected to the Rail Regulator's proposals, but said that the clear understanding was that they were not in any circumstances to be at the expense of safety. Whether there was an effect on safety depended on how they were managed.

9.94   It is unhappy that there was any misunderstanding. More fundamentally, on Mr Coleman's interpretation, it is difficult to see the value of the advice from the HSE. It may be that there are situations in which there is no radical objection on safety grounds but caution is required, in which case one would expect the advice to say so unmistakably. If the situation is that, for whatever reason, the safety regulator has not made an assessment of the risks, the HSE should say so, and may need to advise that the proposals should not be proceeded with meantime. I note that in the letter Mr Coleman said:

> "...HSE has not been able to undertake any very detailed assessment in the time available".

Miss Bacon, apparently in a reference to the same matter, said:

> "We were, I think, being asked to take a view on matters very quickly in circumstances where we probably did not have the resource to deploy as much as we needed on to working through a lot of documentation".

9.95   One may compare this with what was said by Mr Coleman at a liaison meeting on 12 October 2000. He said that the HSE's comments on the safety implications of the efficiency targets which were then contemplated by the Rail Regulator

> "...would be couched as general principles as the HSE were not resourced to conduct a detailed analysis of ORR's proposals".

The Rail Regulator, on the other hand, gave evidence to the Inquiry that he believed the HSE were able to analyse the safety implications of these targets.

9.96   The matters set out in the last three paragraphs demonstrate that, if the HSE are to keep pace with the demands for safety advice which the relationship between the safety regulator and the economic regulator involves, they may need a significant addition to their resources.

9.97   During the Inquiry comparisons were drawn between the powers available to the safety regulator and those available to the Rail Regulator: the former does not enjoy the powers to impose penalties or incentives which are available to the latter, nor does it have the power to impose detailed requirements on Railtrack which he has under Condition 7 of their network licence. I am not persuaded that there is a case for an increase in the powers of the safety regulator. A great deal can be achieved through close co-operation where both regulators want to achieve improvement. Thus, in the case of broken rails, Miss Bacon said that the Rail Regulator took action with the strong encouragement of the HSE. It was not so much a matter of the HSE deferring to the Rail Regulator as

> "...active discussion to make sure that what was done was right from a safety point of view. We were satisfied that it was and we were glad to see action

being taken in circumstances where, frankly, we could not have acted with the same alacrity and effect".

The relationship between the safety regulator and the SRA

9.98 As I stated in para 3.46, under Section 207(3)(a) of the 2000 Act, the SRA are to have regard to, *inter alia*,

> "...the need to protect all persons from dangers arising from the operation of railways (including, in particular, by taking into account any advice given by the Health and Safety Executive)".

Re-franchising

9.99 The policy of the SRA, as I noted in para 3.47, is to use the granting of franchises as a means of driving up safety standards and for enforcing safety compliance by franchisees. A bid by a would-be franchisee is expected to include a safety plan to demonstrate that it had identified the major hazards and how they were to be addressed. This safety plan would appear to be a precis or summary of a safety case. The franchisee was also expected to state targets and how they would be achieved. Mr Morris said that the SRA, in conjunction with Railtrack and the HSE, were to assure themselves that such a safety plan was cogent and in line with accepted principles of safety management. The SRA would seek to ensure that there was a level playing field between different potential franchisees, whether or not they were established operators.

9.100 The SRA's concern with the safety aspects of the bids by franchisees was the subject of discussion at the time when the hearing of Part 2 of the Inquiry was proceeding. At a meeting between representatives of the SRA and the HSE on 23 November 2000 it was agreed, according to the notes of that meeting with which I was provided, that

> "...the SRA will continue to send parts of the replacement bids that relate to safety to the HSE, both at the time of shortlisting and then at the time of preferred bidder. HSE will provide a commentary to the SRA, which will indicate whether the proposals raise any safety concerns, which the SRA will then discuss with the bidder. When the HSE would like to be a party to those discussions, they will be invited to the relevant meeting...HSE's advice to SRA would not rank bidders but would, where necessary, indicate those bidders that were wholly unacceptable on safety grounds".

The minute went on to state that the revised memorandum of understanding would reflect the involvement of the HSE in the re-franchising process, adding:

> "It was agreed that the SRA (in conjunction with the HSE) will use re-franchising to encourage best safety practice, particularly with regard to managing safety from the top and fitment of ERTMS. Thus far, most of the bids have demonstrated a keen awareness of the need for this".

9.101   It may be noted that this followed a letter from Mr Coleman to the Chief Executive of the SRA dated 7 July 2000 in which he emphasised that it was not enough for the SRA to rely on the existence of legal duties and procedures. This was no guarantee of full compliance by duty holders. Moreover there would be differences between applicants in their abilities and approaches in respect of safety management standards. Mr Coleman went on to say:

> "It would I suggest seem odd to the public and Ministers if no assessment of safety aspects, or only a limited assessment without specialist advice, was being made in the re-franchising process".

He pointed out that the process offered an opportunity to ensure that the correct attention was being paid to safety performance and arrangements, to move standards in the right direction and to gain a specific safety payback.

9.102   There is some force to the point which was made by Counsel for the bereaved and injured represented by the Southall and Ladbroke Grove Solicitors' Group, that the HSE should be provided with the whole bids since aspects of commercial operation might have safety implications. Counsel reminded me of the remark by the Rail Regulator that "safety and performance are two sides of the same coin". I do not consider that it is necessary to go so far as to recommend that this should be done in every case. However, the SRA should be alert to a possible need to draw the attention of the HSE to parts of bids, especially in the area of performance, which do not "relate to safety", but may nonetheless raise safety issues (cf para 4.60).

The safety performance of franchisees

9.103   The Inquiry was provided with a copy of the template for the franchise agreements, dated 5 July 2000. In Clause 5.7(3)(a) it stated under the heading of "Safety" that the franchise operator was to use all reasonable endeavours to improve the safety record and safety standards of the franchise services on a continuous basis; submit annually, and whenever else reasonably requested, a plan which would identify specific measurable targets for improvement and would enable it to secure such improvement; comply with such plan to the extent consistent with its obligation to use all reasonable endeavours; and, when and to the extent reasonably requested, produce evidence of improvements in safety records and standards secured pursuant to any such plan. Mr Muttram commented that the conditions of the franchise agreement fundamentally affected the ways in which franchisees would behave in discharging their safety responsibilities.

9.104   Mr Morris explained that in order to ensure that the SRA carried out their duty under Section 207(3)(a), they had invited the HSE to attend the annual appraisal of the performance of franchisees, so that they had the HSE's advice on matters where they did not have enough expertise of their own. The SRA had considered that it was not enough to rely on monitoring by the HSE since they were concerned not only with safety but also with performance as a whole. The SRA had the right in accordance with Clause 21 of the franchise agreement to revoke the franchise in certain circumstances. The clause did not refer specifically to a failure in safety performance. However, it did cover a failure to carry out a commitment such as a safety investment and a case in which the franchisee had been deprived of its licence. In regard to a question of possible revocation the SRA would take the advice of the HSE.

9.105   During the course of the evidence it was pointed out that, in view of the additional safety-related supervisory requirements which the SRA would put in place in regard to operations under a franchise, there could be an imbalance between franchised and non-franchised operations. Mr Morris stated that the point was under discussion in the SRA. They intended to address the situation but would require to think very carefully about how to do so. This issue has not yet been resolved, as far as I am aware.

Safety strategy

9.106   The setting of overall policy for Britain's railways, including an expression of the public's expectation in regard to safety, is, of course, a matter for Government. At the regulatory level, as parties to the Inquiry recognised, there is a need for a statement of strategic objectives, designed to be consistent with Government policy. It is for the rail industry, on the other hand, to set its targets for meeting those objectives and its own performance and safety requirements.

9.107   As I have already noted in para 3.45, the SRA have a duty under Section 206 of the 2000 Act, to formulate, and keep under review, strategies with respect to their purposes. These include, in terms of Section 205, "To secure the development of the railway network". It is expected that the strategy document of the SRA will be issued in the autumn of 2001.

9.108   There was some lack of clarity in the evidence about the relationship between the SRA's strategy and matters of safety. Under cross examination by Counsel for Railtrack Sir Alastair accepted that, using the advice of the safety regulator, the SRA would have "the opportunity and power to take the strategic lead in safety matters", using the advice of the safety regulator "and our position to debate it with others such as RSL (Railway Safety)". In another passage of his evidence he said:

> "We perceive that we have to guide, lead, encourage, push the industry in the direction of safety, yes. What constitutes safety is what we are asking for somebody, a specialised agency, to establish".

The Rail Regulator, speaking of the SRA's strategic role, said that "safety is at the very heart of the strategies that it is going to have".

9.109   However, other evidence suggested a more modest role for the SRA in regard to the matter of safety. In their statement of case they stated that they intended

> "...to formulate, having taken into account particularly the advice of the HSE and others, strategies which recognise the need to introduce measures which have as their aim the amelioration of current major safety hazards applicable to railways".

In his evidence Mr Morris said:

"We have inferred...that we do have a duty within the Strategic Rail Authority to have positive regard for safety issues in the whole range of duties that we undertake. What we have done is to try and put into our safety strategy document the manner in which we would exercise those functions".

He correctly accepted that the SRA had a safety duty as opposed to a safety function. According to the notes, at the meeting on 23 November 2000 to which I referred in para 9.100,

"SRA confirmed that HSE would be consulted on SRA's strategy document. This document will not contain a Safety Strategy for the rail industry but will set out how SRA plans to discharge its safety responsibilities. There was general agreement that a safety strategy for the rail industry was needed, and that consulting on and setting such a strategy fell to HSC/E".

No such safety strategy had been produced, at least to the knowledge of the Inquiry, prior to the closing of the hearings in Part 2 on 20 December 2000.

9.110   In the course of her evidence Miss Bacon emphasised the role of the HSC. "I do think", she said,

"...that HSC needs to be setting a strategic framework of regulation and safety policy, and issuing guidance, taking a view on best practice and so on".

She said that in response to the higher profile which was being given to regulatory activity over the previous 12-18 months, more was being expected of the HSC.

9.111   Putting these views together it is reasonably clear, and appropriate, that the strategy for which the SRA are responsible should contain a safety element, in respect that their strategy for the development of the railways and other matters has to have regard to matters of safety. For this purpose they should look to the safety regulator for advice. On the other hand, safety strategy as such is and should be a matter for the safety regulator, specifically the HSE in their role as the body for providing guidance on safety policy at the regulatory level. It is perhaps unfortunate that the inter-relationship between the functions of the SRA and the HSE in regard to matters of strategy is not spelt out in the statute.

9.112   Among other witnesses Miss Bacon emphasised the need for better co-ordination between the safety regulator, the Rail Regulator and the SRA, stating:

"This is not just joined up working. It is getting the sequence of events on investment decisions, franchising decisions, access conditions, what the safety regulator, HSC, is trying to achieve in overall strategic terms and thinks to be reasonably practicable, etc".

Chapter 10
A rail industry safety body

Introduction

10.1 So far in this report I have reached the view:

    (i)   that the function of setting Railway Group Standards should be assumed by a body independent of Railtrack Group plc and their subsidiaries (para 8.38); and

    (ii)   that it is inappropriate that the safety regulator should perform this function (para 9.46). This points clearly to the conclusion that in principle the function should be assumed by a new rail industry safety body which is independent of both Railtrack and the safety regulator.

10.2 Starting from that conclusion I discuss in this chapter the role, functions, constitution and resourcing of such a body. It deals with the following subjects:

- the models proposed by parties (para 10.3);
- the setting of standards (paras 10.4-10.10);
- auditing and enforcement (paras 10.11-10.14);
- accrediting and licensing (paras 10.15-10.16);
- approval of rail vehicles (paras 10.17-10.20);
- constitution (paras 10.21-10.27);
- organising for safety (paras 10.28-10.29); and
- European regulation (para 10.30).

The models proposed by parties

10.3 Models for a new rail industry standards body were proposed in this Inquiry by the Joint Rail Unions, the Collins Passengers' Group, the Rail Regulator and the HSE. As can be seen from Appendix 5, these models differed from each other in a number of respects, in particular whether the body would also have the function of ensuring compliance with Railway Group Standards; whether it would have other functions; and whether it would be established as a statutory authority. These are some of the questions which I will discuss in the course of this chapter.

The setting of standards

10.4 As I mentioned in para 9.69, in their report to the DG Energy and Transport, NERA described the concept of a national body representing the infrastructure and train operators and the supply/maintenance industry for the purpose of setting standards and decision rules. In their view, the safety regulator's role should be to supervise this and not to perform it. Such a body, they considered, would also ensure that adequate research and development on system-wide safety issues were undertaken, and promote a forward-looking approach, for example on national policy towards European harmonisation. The Inquiry was informed that Canada is an example of a country in which the rules for the industry are written for it by an industry body.

10.5   Sir David Davies, President of the Royal Academy of Engineering, made a number of important points in regard to a rail industry standards body. He said that it was feasible but it would have to be consistent with Railtrack's interest in the control of the importation of risk on to the infrastructure. It should command the confidence of the industry. It had to have power to make binding decisions. There required to be a mechanism for the fair resolution of disputes. It should be perceived to be independent.

10.6   As regards the standards which would fall within the province of the body, they should, in my view, include what are presently administered by Railway Safety as Railway Group Standards. While, as I understand the matter, all of these standards have some safety content, many are to a greater or lesser extent concerned with non-safety matters. I do not favour the distinction which the Collins Passengers' Group sought to draw, based on the evidence of Dr Cox, between "technical rules and procedures" and safety regulations, or likewise their proposal that the new body should incorporate the technical division of the HMRI (Division RI1). Apart from cutting across a well-recognised division of responsibilities, the distinction which they sought to draw does not seem to me to be one which can readily be put into practice. Furthermore, as Mr R I Muttram, Director of the S&SD, observed, it does not clearly distinguish between the technical rules which apply to a safe system and safe interworking and those which are applied by individual companies such as Railtrack Line. This is an important distinction (see paras 6.14-6.15).

10.7   The setting up of the body also provides the opportunity to bring within its province standards of the type which have ceased to be Railway Group Standards in respect that they are concerned only with the interior of rail vehicles. In this connection I noted that, even as matters presently stand, Sir David pointed out that there is a need for standards across the whole industry, referring to the agreement between Railway Safety and ATOC for the taking on of additional standards (see para 3.55). He was also plainly conscious that there was a need for the rail industry to have "more ownership" of the actions of Railway Safety.

10.8   The body should be advised by the SAB and serviced by the RISSC and the Standards Sub-Committees in the same way as Railway Safety is at present. There does not appear to be any need for any radical change in the membership of these bodies. However, it is for consideration whether in due course:

(i)   the activity of the body which I recommend will make the separate existence of the SAB unnecessary; and

(ii)   the RISSC should become a strategy committee of the body.

I should add that it would also be appropriate that the body should be responsible for the preparation of any proposed changes to the Railway Group Standards Code.

10.9 The body should have explicit duties to set and review standards. The performance of its duties should be subject to the supervision of the HSE through auditing and other actions. I agree with the proposal of the Rail Regulator in his statement of case that the role of the safety regulator should be to ensure that the standards which are set meet the current requirements both of safety legislation and of good practice. It would thus have a supervisory function and an obligation to "prod" the body into action when standards had fallen behind current requirements.

10.10 The standards for which the body has responsibility should be binding not merely on members of the Railway Group but on any company to which the requirement to comply currently applies, whether by virtue of a licence condition or a contractual term. This should not be taken as qualifying the duties of Railtrack under health and safety legislation, or their ability to do what is necessary for the performance of those duties (cf para 8.38).

Auditing and enforcement

10.11 Unlike the three other proponents of a rail industry standards body, the Collins Passengers' Group advocated that the body should also audit the compliance of companies with standards.

10.12 However if it had such an audit function, it would mean that operators would be subject to external auditing by three separate bodies - the others being Railtrack and the HSE. This seems to me to involve an unnecessary duplication. I noted that Mr Coleman advocated better and more proactive audits. He said that one improved basic audit could remove the necessity for other audits. I agree with that general approach.

10.13 As regards external auditing by Railtrack, this could incorporate the audit function previously discharged by Railway Safety. The escalation procedure which I described in para 7.41 would apply. A greater burden of auditing would fall on the HSE, but this would be consistent with their departure from the previous "light touch" approach.

10.14 At the same time it is important, I consider, that the rail industry body should have the benefit of feedback from the auditing carried out by Railtrack and the HSE, in order to assist it in considering improvements to existing standards as well as the setting of new ones. Non-compliance may suggest that an existing standard is in need of clarification or replacement.

Accrediting and licensing

10.15 I am in no doubt that if the industry standards body is to be set up it should also be responsible for what I have referred to in Chapter 7 as the accrediting of suppliers of products and services and the licensing of individuals. Once again this would be subject to the supervisory activity of the safety regulator.

10.16 As matters develop, it may be opportune for the industry body to move into a closer relationship with other bodies such as the Institution of Railway Signal Engineers (IRSE) and the IRO.

Approval of rail vehicles

10.17 It is highly desirable that steps should be taken to streamline the processes for the approval of new rail vehicles. This is an area where I would expect the industry body to take an active role in the achievement of improvements.

10.18 At present there appear to be two matters of concern. The first is in regard to the time taken by the processes under the existing Railway Group Standards, and in particular that relating to route acceptance, which is under the direct control of Railtrack. The Inquiry heard evidence of complaints by ROSCOs and the RIA, with unfavourable comparisons being drawn between experience in Great Britain and that in other European countries. The NERA report observed:

> "The Route Acceptance process was in respondents' view a major source of cost and distraction, imposing risks (sometimes realised) of severe delays, with negligible safety benefit and possibly the reverse (because of delays in introducing safer equipment)".

Mr Muttram said that the complexity was inevitable: it was due to the complexity of the network which had been built over a long period and had a great deal of variation. It appears that the principal cause of difficulty is concern about the possibility of electrical interference. Railtrack said that they had spent a lot of money on the collation of all the relevant information about routes with accessible databases. However, ROSCOs reported that considerable progress had been made in various meetings of the SRA's Working Group on Vehicle Acceptance and Maintenance. Meanwhile the Rail Regulator has proposed a modification of Railtrack's network licence requiring them to establish a register of the condition, capability and capacity of their assets.

10.19 Another aspect is the duplication of functions as between these processes and that carried out by the HSE, even allowing for what was said about its limited nature. As I have already noted, this is presently the subject of consultation (see para 9.57).

10.20 I should add that I do not consider that it is appropriate for me to recommend an alteration to the arrangements for the administration of the Rail Vehicle Accessibility Regulations (see para 3.58), which are concerned not only with safety but also with other aspects of the facilities for the disabled.

Constitution

10.21 The body should, in my view, be set up as a new legal entity, independent of any company in the rail industry and of any part of that industry, such as, for example the TOCs. It should have the power and the duty to take binding decisions.

10.22 I envisage that the arrangements for the governance of the body would make provision for the representation of railway operators (i.e. Railtrack and the train and station operators), and of any other company to which the requirement to comply with Railway Group Standards (or the additional standards to which I referred in para 10.7) applies, whether by virtue of a licence condition or a contractual term. There should also be representation of the manufacturers and suppliers of infrastructure equipment and rolling stock. Since the number of bodies in any given category may be numerous, some form of collective representation, for example by ATOC and the RIA, would be necessary. I also consider that there should be representation of the three main rail trade unions. As regards representatives of the travelling public, the best way in which they can make, and be seen to make, an effective contribution is at a higher level (cf para 6.17).

10.23 One of the points to which Sir David Davies correctly drew attention was the need, in the case of a body which is drawn from an industry where members have divergent interests, to avoid frustration of its purpose through impasse or conflict. Further it is essential to avoid a situation in which the development of Group Standards is dependent on consensus: this could lead to a level of standard which represented no better than the lowest common denominator. For these reasons, if for no others, I favour the appointment of an independent chairman and provision for a number of independent members. They should have suitable practical experience. There should be a clear and easily accessible means of resolving any matter which is in dispute. This should be taken first to the body and thereafter, if necessary, the decision of the Rail Regulator, acting with the advice of the HSE as he would do under the present arrangement.

10.24 It was submitted on behalf of the Collins Passengers' Group that the body should be set up under statute in order to give it "teeth" and ensure that it had and commanded confidence. The first of these two points is obviously based to some extent on the question of auditing and enforcement, with which I have dealt above. For the HSE it was contended that a statutory basis would tend to detract from the importance of the railway operators fulfilling their duties under the 1974 Act.

10.25 While there are some attractions in a statutory basis, I am not convinced that it is the most appropriate way to proceed. Quite apart from the point made by the HSE, which has some merit, I have a concern that a statutory body would be seen as a "quasi-regulator". It is important for public confidence that the regulation of safety should be seen to be unambiguously in the hands of one body.

10.26 As to a non-statutory basis, one possibility which I consider worthy of investigation is that the body be constituted by a means of a modification of Railtrack's network licence and the licences of the other railway operators. Such a modification could be used not only to bind the individual licence holder to comply with technical standards, but also to require it to bind the companies with which it contracts. The modification could also cover the setting up, functions and supervision of the body, the arrangements for participation in its work, the binding nature of its decisions and the means of resolving disputes. It appears that at present the Rail Regulator does not have the power to require licences to be modified to this effect. If so, an amendment of the relevant transport legislation would be required.

10.27 I envisage that the Rail Regulator would be an appropriate person to undertake the setting up of the body. It should be funded, in my view, by means of a levy on the companies covered by the requirements referred to in para 10.22.

Organising for safety

10.28   The creation of an industry body for the purposes I have outlined earlier in this chapter provides, in my view, an appropriate base on which to build other functions which are truly matters for the industry. I mentioned a number of them in paras 9.63-9.64, namely:

      (i)   establishing and managing system authorities;

      (ii)   funding and sponsoring research and development;

      (iii) monitoring and reporting on the industry's safety performance; and

      (iv) the development of the annual Railway Group Safety Plan.

Another is the dissemination of good practice. Last, but not least, is safety leadership.

10.29   The creation of the body I have been discussing seems to me to provide an excellent opportunity to re-create part of what was lost as a result of the disaggregation of the industry. As I stated in para 9.6, it seems to me that the most productive approach is to ask what can be done to enable safety to be as effectively managed and regulated as it would be if the industry were a single enterprise. A rail industry safety body offers a clearly defined way in which the rail industry can collaborate in the promotion of safety in a way which cannot be achieved by any one member of the industry acting on its own. I refer to my remarks in paras 5.17 and 5.67-5.68.

European regulation

10.30   What I have stated above in regard to the setting of standards such as Railway Group Standards should, of course, be read along with what I have said in paras 9.67-9.74 about European regulation. I should, however, add that, even if it proved to be the case that responsibility for the setting of such standards is required by the Directive in its final form to be taken over by the safety regulator, it is preferable that a new rail industry safety body should be set up and assume the other functions which I have recommended in this chapter.

Chapter 11
An accident investigation body

Introduction

11.1   The main purposes of this chapter are to discuss:

      (i)   the investigation of accidents and incidents under the RAIB which I recommended in para 9.29; and

      (ii)   the relationship between such investigations and those carried out by the police.

11.2   The main witnesses from whom the Inquiry heard evidence on these matters were Dr M H Walter, Controller of Safety Management Systems in the S&SD; Ms A E Forster, Operations and Safety Director for FGW; and Mr D J Williams, Chief Constable of the BTP.

11.3   The chapter will accordingly cover the following subjects:

- the investigation of accidents and incidents (paras 11.4-11.22); and
- the relationship with police investigations (paras 11.23-11.33).

The investigation of accidents and incidents

The status quo

11.4   Before coming to the discussion of what is proposed for the future, it may be helpful to take into account some features of the present arrangements. An accident may be the subject of an "investigation" under Section 14(2)(a) of the 1974 Act, which is normally carried out by the HSE under the direction of the HSC. In para 9.26 I have already set out a number of criticisms which were made of such investigations.

11.5   Accidents and incidents may also be the subject of a formal inquiry or formal investigation under Railway Group Standard GO/RT 3434/3. Currently about 90 and 150 respectively are arranged per annum. According to the Group Standard, a formal inquiry denotes a formally structured inquiry generally implemented in the case of high potential or major accidents. It is to be held in cases of accidental death; in circumstances involving accidental or multiple injuries resulting from a serious train accident, or other accidents "where a Public Inquiry or HMRI Inquiry is likely"; or where the circumstances are such that it is considered to be necessary to ensure the facts are fully investigated. The inquiry is carried out by a panel representing the organisations involved in the accident. A formal investigation is held where this is necessary and a formal inquiry is not being held. A formal investigation is carried out by a person or team selected by the lead organisation involved in the accident. Formal inquiries and formal investigations have as their objectives to:

"Establish the full facts; determine the immediate and root cause(s); assess compliance with Railway Group Standards; question whether methods of working are safe; determine whether specific actions are necessary to avoid recurrence; determining *(sic)* whether changes are necessary to training, supervision, instructions, maintenance schedules, equipment used, etc; question whether there are underlying weaknesses, e.g. in the organisation, safety management systems and associated controls; enable prevention of recurrence".

Regulation 11 of the Railways (Safety Case) Regulations 2000 places a duty on railway operators to co-operate with the holders of safety cases to enable the holders to comply with them. Their safety cases have to describe their arrangements for the investigation of accidents and incidents. A duty of co-operation is also imposed by Regulation 11 of the Management of Health and Safety at Work Regulations 1999.

11.6   Dr Walter informed the Inquiry that in June 1998 he had been commissioned to ascertain what was needed in order to ensure that formal inquiries were re-focused to deal with major issues, determine underlying causes and deliver objective recommendations. This led to the S&SD consulting Railway Group members and other stakeholders by a Green Paper issued in May 1999. A year later the S&SD issued a White Paper which set out the responses along with the proposals of the S&SD. In May 2000 Dr Walter produced a further paper "Accident Investigation – Proposals for Change", which was submitted to the Railtrack Group Committee. This is the day to day business committee of the Boards of Railtrack Group plc and Railtrack plc, which among other things considers urgent or important safety issues. The committee endorsed a number of proposals in the paper, the implementation of which is under way. Dr Walter said that the process had identified certain weaknesses in investigation. There was a perception of a lack of independence and objectivity; a failure by the industry to respond effectively to the recommendations of inquiries; a tension between criminal and industry investigations which affected the ability readily to reach conclusions on root and underlying causes; a fear of prosecution hampering the free and open recognition of error and hence the proper learning of lessons; and the need to improve the standard of competence of investigations and the focus on recommendations.

11.7   As I have stated above, the Group Standard requires that the panel conducting the formal inquiry should represent the organisations involved. Ms Forster was critical of this requirement which she regarded as "unsustainable". She added:

"I think it has been done with great integrity, but I think it is not right. It is not seen to be right. It cannot be accepted by others looking on as an open and honest process".

In his report on the Southall crash Professor Uff's Recommendation 78 was that the panel should be independent of all parties having an interest in the accident. Ms Forster accepted that in the case of formal investigations it would suffice that there should be an independent input, depending on the nature of the case. She also advocated that there should be greater transparency of the process, saying:

"I think what is vital, there is little confidence in the way that the industry has kept things behind closed doors. Clearly that must change and confidence will be gained through experience and by making it a much more transparent process to the public".

She also advocated a process for the challenging of a finding. She gave as an example a case in which the Safety Review Group in a Railtrack Zone had disputed the conclusions of the panel and asked that it sit again to review its decision but without hearing further evidence. She pointed out that there was no mechanism by which the original conclusion could be verified or changed.

The scope of responsibility of the RAIB

11.8   There was substantial support for the view that the investigation of accidents and incidents of whatever nature should be brought under the overall control of the RAIB. I will, for convenience, refer to them as "cases". Accordingly all cases would require to be reported to the RAIB. The more serious cases would be the subject of what I will refer to for convenience as inquiry by the RAIB itself. The less serious would be delegated to the industry to be dealt with by formal inquiry or formal investigation. However the RAIB would have the ability to call in any case for inquiry by itself where that appeared to be appropriate. An example might be where a series of events suggested the risk of a more serious repetition. Formal inquiries and formal investigation by the industry would be under the general supervision and guidance of the RAIB.

11.9   A benefit of this system is that it would eliminate the existing overlap in investigation between the safety regulator and the industry. It would also provide a single means of providing supervision and guidance. It would not, of course, eliminate the need for individual companies to carry out their own internal investigations where they considered that necessary for their own purposes. I recommend the adoption of such a system. I also endorse the suggestion that industry parties should be alert to the need to invite the RAIB to inquire into cases where they consider that this is appropriate. It was pointed out in the course of the evidence that the effect of this system was to reduce greatly the need for formal inquiries, since the more serious cases would be inquired into by the RAIB. This suggested that it might be appropriate for there to be only a single method by which accidents or incidents were investigated at industry level. I can see the merit of that suggestion, but at this stage it is difficult to predict to what extent there will be a remaining need for formal inquiries. However, this is a matter which should be considered in the longer term.

11.10   There was some discussion as to the categories of case which would fall to the RAIB to inquire into, apart from any other cases it had called in. I do not consider that it would be wise or useful for me to draw up a list of categories, since it would be better for this to be the subject of study. Dr Walter stated that the S&SD had considered the categorisation used in a number of other countries, including the United States of America. Reference was made to the definition of "significant accidents", which fall to be investigated by the NTSB. He regarded this formulation as a good working approach, subject to the need to take account of "near misses", the significance of which should never be understated.

11.11 The evidence before the Inquiry plainly supported the view that inquiries by and under the RAIB should concentrate on the search for root causes rather than to ascribe fault, and that the investigation process should not be distorted by questions of civil liability or criminal responsibility. This is, of course, the general approach taken by the AAIB and the MAIB. Regulation 4 of the Civil Aviation (Investigation of Air Accidents and Incidents) Regulations 1996 states:

> "The sole objective of the investigation of an accident or incident shall be the prevention of accidents and incidents. It will not be the purpose of such investigation to apportion blame or liability".

11.12 It may be noted that, while the Group Standard states that one of the objectives of a formal inquiry or a formal investigation is to determine root causes, the report by HSE "The Management of Safety in Railtrack" remarked that it lacked

> "…a definition of an investigative process which will establish underlying causes of failures which led to the accident".

The independence of the chairman and panel members in formal inquiries

11.13 Dr Walter accepted the desirability of an independent chairman in formal inquiries. Provision is made for such a chairman in Group Standard GO/RT 3434/3. He also agreed that there was merit in the suggestion that in the most serious cases coming before a formal inquiry, the members should come from a different Railtrack Zone and TOC from those with which the case was concerned. He had no doubt that this was already the case. For the industry inquiry into the Ladbroke Grove crash there had been an independent chairman; and in the case of the Hatfield crash both the chairman and the other members of the panel were independent. I endorse this general approach, which depends on the circumstances of the case. It is highly desirable that there should be a clear and publicised statement as to the practice that is to be followed as to the independence of the chairman and panel members.

Attendance at inquiries

11.14 During the Inquiry there was some discussion about whether representatives of those who had been bereaved or injured in an accident should be allowed to attend formal inquiries. In the White Paper issued by the S&SD the view was expressed that to allow members of the public or their representatives to attend formal inquiries would probably obstruct rather than assist their main purpose. However, it was noted that representatives of the Rail Passengers' Council were allowed to attend the formal inquiry into the Hatfield crash. Dr Walter said that the S&SD had very much an open mind and could see great benefits in the involvement of public representatives. He said that the passage in the White Paper had been written with reference not only to accidents which had attracted great public interest but also to others where it was useful to maintain a relatively small inquiry in order to avoid witnesses being deterred. He added: "But, in essence, I think certainly openness is deep in our thoughts". I note that Recommendation 79 by Professor Uff was that consideration should be given to whether procedures could be adapted to make any industry inquiry accessible to the public, save where the needs of confidentiality otherwise required. I can well see the argument in favour of allowing representatives of those who have been affected by an accident to be present as observers in the case of a formal inquiry into a relatively serious incident. I consider that, save and to the extent that there is good reason to the contrary, representatives of those affected by an accident should be allowed to attend as observers. As Dr Walter accepted, there is a need for a criterion for determining for which inquiries this would or would not be suitable. There was little discussion in regard to inquiries which would be conducted by the RAIB if that body is set up. However, I agree with the submission made by Counsel for the bereaved and injured represented by the Southall and Ladbroke Grove Solicitors' Group that such representatives should be allowed to attend such inquiries as observers.

11.15   I fully agree with the submission made by ATOC that it is essential that those who have been bereaved or injured are kept fully informed of what is happening during the investigation process. I agree that the AAIB procedures, which I have outlined in Appendix 6, are exemplary.

Supervision

11.16   The RAIB should, in my view, exercise a supervisory function in regard to the working of formal inquiries and investigations. Thus, for example, the RAIB should issue guidance in regard to good practice, and monitor the working of the system in the industry. I also consider that there is some merit in the suggestion made by Ms Forster that the RAIB should provide a means of reviewing a finding or recommendation of an industry inquiry which has been disputed. It would, of course, be preferable that such a situation does not arise: if there is some challenge it should, where possible, be raised in the inquiry before the decision is reached. However, I can envisage a situation in which fresh information came to light after the decision. This proposal should be the subject of further study. It is obviously essential that any appellate process should be conducted with both fairness and dispatch.

The publication of reports of investigations

11.17   There was clear support in the evidence for the publication of the reports of RAIB inquiries and formal inquiries. It was pointed out on behalf of the Joint Rail Unions that the ETSC had recommended that staff should be advised of the results of accident investigations and new safety recommendations. It was also in accordance with the practice of the AAIB (see para 20 of Appendix 6). The only matter of concern was that the identity of persons involved should be protected, in the same way as that followed by the AAIB. I support this approach.

The following up of recommendations

11.18   In the light of the evidence in the Inquiry it is clearly necessary that there should be an effective system for the following up of recommendations arising out of RAIB inquiries and industry inquiries.

11.19   At present Railway Group Standard GO/RT 3434/3 requires members of the Railway Group to have procedures for responding to, and implementing, recommendations. However, it appears that external monitoring has not been effective. The HSE's report

"The Management of Safety in Railtrack" found that there was little evidence that Railtrack were reviewing the progress of TOCs in the implementation of the recommendations of formal inquiries. It was also found that Railtrack were unable to demonstrate that the actions which had been placed on TOCs had been tracked to completion. In the S&SD paper "Accident Investigation – Proposals for Change" it was proposed that each organisation to which a recommendation had been directed would be required, *inter alia*, to provide the S&SD with:

(i)   full details of the action that it had taken or proposed to take, or a full justification for not doing so; and

(ii) progress reports to required timescales until close-out of the action.

It was also proposed that the S&SD would maintain a database of all recommendations from formal inquiries and the responses of individual organisations, including Railtrack Zones, and would evaluate action plans for fitness for purpose and, through receipt of progress reports and sample audits, monitor actions to their close-out.

11.20   I am in no doubt that a system of the type which I have outlined above is required in order to ensure that all organisations to which recommendations made by RAIB and formal inquiries are directed proceed with due expedition to implement them, and that, if they do not do so, it is for good reasons.

11.21   In my view the appropriate body which should maintain a current record of these matters is the rail industry safety body which I recommended in Chapter 10.

11.22   Because formal investigations are generally concerned with matters of less moment, I do not consider that it is practical or necessary for a national record to be kept of their recommendations. However, the RAIB should regularly examine their reports in order to determine whether there are matters of local or national importance which should be brought to the attention of the industry, for example if a number of accidents suggested a significant trend.

The relationship with police investigations

11.23   The BTP provide policing services to the national rail network, including the protection of members of the public and employees. The interests of the British Railways Board, who were the employers of the force, were transferred to the SRA under the 2000 Act. The intention of the Government is to introduce legislation in due course to create a BTP Authority which would become the employer of the force.

11.24   By convention the BTP carry out the police investigation of rail accidents. It is, of course, the case that only a relatively small proportion of accidents are subject to such investigation.

11.25   It was common ground in the Inquiry that there is an overriding interest in the swift determination of the causes of rail accidents, the publication of the report and the implementation of any safety lessons. It was also common ground that it was essential that safety-critical information which is gleaned in the course of the investigation of an accident should be made available forthwith to those who had the responsibility of management of safety on the railways.

11.26 BTP rightly submitted that, while the possibility of criminal proceedings was very important, it should not interfere with this overriding interest, save in exceptional circumstances. As to the latter they instanced a case in which the facts revealed the commission of a very serious offence by one of those bodies which would normally be entitled to free access to information obtained from the investigation. In such circumstances the public interest might require the temporary reservation of some non-safety-critical evidence.

11.27 Where an accident requires to be the subject of a police investigation at the same time as an investigation carried out by a rail accident investigatory body, there appears to be no good reason why both investigations should not proceed in parallel. Arrangements for this purpose seem to have worked well in the case of the crash at Ladbroke Grove. The position of the BTP, as set out in the Inquiry, is that while they are the only body which is capable of large scale investigation and the swift amassing of large amounts of evidence, they cannot do this alone. As Chief Constable Williams put it:

"Where technical evidence is in issue, whether it be the concept of the RAIB, the current HMRI or whatever, these are the people who should decide on issues of technical nature. BTP have no expertise in this direction".

Accordingly the BTP see a need to work in partnership with the agency which has the necessary technical expertise.

11.28 The evidence at the Inquiry clearly pointed to difficulties arising from the fact that there is a limited pool of independent technical experts, and the fact that a large number of them are normally under contract to a railway operator, such as Railtrack.

11.29 I am satisfied that, as was contended by the BTP, there is a need for a protocol dealing with the release of technical information and access to technical experts. This would cover the safeguarding of the industry's legitimate need for technical information in the aftermath of a rail accident, and the right to call on the services of expert witnesses who are normally under contract to railway operators. At a late stage of the Inquiry the BTP produced a draft model protocol. This included, amongst other things, provision for a joint investigation team.

Statements of witnesses

11.30 There was general agreement that evidence of the statements of witnesses which was obtained in the course of an investigation by the RAIB should, in general, not be disclosed to the police. Chief Constable Williams stated in evidence that, since the overriding purpose was to find out the causes of accidents and use the results of the investigation to prevent future accidents, an RAIB investigation should be a "no blame" investigation. Accordingly it should be "ring-fenced", in the same way as the investigations by the AAIB and the MAIB. The main benefit of this is that witnesses would feel able to speak freely in such an investigation.

11.31 At present an HMRI inspector has the power under Section 20(2) of the 1974 Act to require any person whom he has reasonable cause to believe to be able to give any information relevant to any examination or investigation which the inspector is making to answer such questions as the inspector thinks fit to ask and to sign a declaration of the truth of his answers. Under Sub-section (7) no answer given by such a person in pursuance of such a requirement is to be admissible in evidence against that person or the husband or wife of that person in any proceedings.

11.32 Counsel for ATOC submitted that the statements made by witnesses in connection with RAIB inquiries and industry inquiries and investigations should not be disclosed to the police save by order of a judge. It was submitted that in this way the requirements of public safety would be fulfilled, while the judge could balance competing interests, giving appropriate weight to any human rights issues. Accordingly in ordinary circumstances the police would simply have to take a fresh statement. Such an arrangement is similar to that applying in the case of the AAIB (reference may be made to Regulation 18 of the Civil Aviation (Investigation of Air Accidents and Incidents) Regulations 1996, read with para 5.12 of Annex 13 to the Chicago Convention on International Civil Aviation 1944).

11.33 I endorse this proposal, which, in my view, should contribute significantly to disclosure of information which will enable accidents to be swiftly investigated and their lessons learned.

Chapter 12
Summary of recommendations

Introduction

12.1 In this chapter I will set out my recommendations in the light of the matters which I have discussed earlier in this report. I thank the representatives of the parties, along with members of the public, for their suggested recommendations, all of which I have considered.

12.2 Each recommendation is followed by a reference to the paragraph in the report to which it is most directly related. I set out the recommendations in, so far as possible, the order of the paragraphs to which they refer.

12.3 While I attach importance to all of my recommendations I regard some of them as central. These key recommendations are printed in italics.

12.4 Beside each recommendation is the name of the body (in some instances more than one) which should, in my view, be primarily responsible for its implementation.

12.5 I have given consideration to the period which I should recommend for the implementation of each of my recommendations. I have sought to avoid the imposition of unrealistic requirements. The periods which I have selected are denoted by numbers, as follows:

- '1' means up to 6 months
- '2' means up to 12 months
- '3' means up to 3 years

In some cases, as will be obvious, the recommendation refers not only to initial implementation, but also to continuing implementation thereafter. I have not stated a period where that is unnecessary or inappropriate, as in the case of work that is presently ongoing. I draw particular attention to the words "up to". All reasonable efforts should be made to achieve implementation in significantly less than these periods.

Interfaces and the number of franchises

1.	Railtrack and ATOC should work jointly with the RITC to set up a task force for ensuring that the need for a skilled and properly trained workforce at all levels of the industry is met (para 4.35).	Railtrack ATOC RITC 1

Large scale projects and the case for system authorities

2.	The arrangements for the establishment of system authorities should ensure that they are properly empowered, provide clear leadership and command the commitment of all parties to their work and decisions. System authorities require the means of enforcing their decisions. They should have adequate finances, through proper and equitable contributions from participating bodies (para 4.48).	Railtrack ATOC

Research and development

3.	Subject to Recommendation 55, research and development should, as matters stand, be led by Railway Safety but with the support of the SRA and the Rail Regulator. Further funding should be based on a levy on the participating bodies in proportion to their railway-based income (para 4.54).	Railway Safety SRA Rail Regulator

The use of contractors

4.	*Steps should be put in place to ensure that contractors and sub-contractors are selected by a process which gives due regard to their state of training. They should be given appropriate time further to develop their training and planning as necessary before embarking on work (para 4.72).*	Railtrack IMCs TRCs 1
5.	Steps should be taken to ensure that the quality of work carried out by contractors and sub-contractors entirely meets the required standards, and that any deficiencies are addressed in a timely manner (para 4.75).	Railtrack IMCs TRCs 1
6.	*The Sentinel system should be reinforced with specific reference to the need to record the total hours that any individual works on the railways, and to ensure that the Sentinel card is clearly "tied" to an individual (para 4.80).*	Railtrack IMCs TRCs 1

7.	The steps taken to reduce the number of sub-contractors are endorsed (para 4.82).	
8.	*The taking by Railtrack of a direct and active role in the close day to day management of safety-critical work is endorsed (para 4.83).*	
9.	*Employers of contractors and sub-contractors should ensure that they work to exactly the same safety standards as those who are directly employed (para 4.87).*	Railtrack IMCs TRCs 1
10.	The proposal of a training school for contractor staff is endorsed (para 4.87).	

The role of the trade unions

11.	Management should ensure that the elected representatives of the employees, whether they be union members or not, have a significant role in the management of safety (para 4.99).	Rail industry

Safety leadership within individual companies

12.	*The Chairmen and Chief Executives of companies should make continually clear to all their employees and passengers a lasting commitment to improve safety performance (para 5.21).*	Rail industry .
13.	Companies in the rail industry should be expected to demonstrate that they have, and implement, a system to ensure that senior management spend an adequate amount of time, devoted to safety issues, with front line staff (para 5.23).	Rail industry 1
14.	*Where it is not already in place, a safety management strategic leadership team should be established in each company in the rail industry. Such a team should be led by the Chief Executive and include his or her direct reports, with support from the safety professionals. It should consider the strategic management process for safety by holding regular meetings devoted to health and safety issues. It should be the key group in the organisation for setting goals, monitoring performance and assessing and resourcing the needs of the organisation to ensure that the long-term objectives are met (para 5.24).*	Rail industry 1

Communications

15. *Safety meetings should be used as a means of two-way communication between management and the workplace, and should be linked directly to safety management leadership teams referred to in the previous recommendation (para 5.34).*

Rail industry 1

Risk assessment

16. The greater use of risk assessment in the rail industry is commended (para 5.42).

Rail industry

Railway Group Standards

17. There should be a systematic review of the standard setting process to assess whether it is effective in achieving its overall aim of safe interworking (para 6.18).

Railway Safety RISB

Safety cases

18. The application of the safety case to Great Britain's railways is endorsed (para 7.9).

19. The definition of responsibilities for the control of risk at specific sites which are shared by different railway operators and at the interfaces between them across the network should be refined and set out in the safety case. However, the details of the arrangements and agreements for these purposes should not be required to be set out in the safety case; it should be sufficient that the safety case provides information as to the means of access to them (para 7.15).

Railtrack TOCs 2

20. *A duty holder should be required to show by means of its safety case that it has reduced the risks associated with its operation as low as reasonably practicable, but it should be sufficient if the safety case points to the methods which have been used and to where the details can be found (para 7.20).*

HSC Duty holders 2

21. *Duty holders should be under a statutory duty to comply with Railway Group Standards in so far as they relate to matters of health and safety (para 7.22).*

HSC

22.	The process of safety case acceptance should include the check that a system as described in the safety case is actually in place; whereas the audit would concentrate on how that system was working in practice and how it was ensuring and improving safety (para 7.30).	HSE Railtrack
23.	*It is essential that companies operate a robust internal audit system, which should be both "top down" and "bottom up" (para 7.37).*	Rail industry

The accreditation of suppliers and producers of services

24.	*Suppliers of products or services of a safety-critical kind for use on, or in regard to, the railways in Great Britain should be required to hold an accreditation as a condition of being able to engage in that activity. But the features of such a system require further study (para 7.73).*	HSC 3

Licensing

25.	There should be a system for the licensing and central recording of those who are qualified for the driving of trains in respect of their knowledge of the rules and regulations and the traction for which they have been assessed as competent. Training providers or train operators should be accredited and common standards laid down for the purpose. Drivers' licences should require to be revalidated every three years (para 7.74).	RISB 3
26.	There should be a similar system for licensing the central recording of qualified signalmen, based on an assessment of their knowledge of the rules and regulations. Revalidation every three years should be required (para 7.75).	RISB 3

Railtrack and Railway Safety

27.	*The transfer from Railtrack to the safety regulator of the function of acceptance of the safety cases of train operators and station operators (and their material revisions), and the removal from the S&SD of their function in regard to safety cases and Group Standards, are endorsed (para 8.28).*

28.	*The safety regulator should cease to be dependent on Railtrack for a recommendation as to whether or not the safety case of a train operator or a station operator (or its material revisions) should be accepted. Instead the safety regulator should give Railtrack the opportunity to make any representation as to whether or not the safety case or revision should be accepted, and the grounds on which such a representation is based. The safety regulator should likewise give the opportunity to any other train operator or station operator who may be affected by matters referred to in the safety case to make a similar representation, and for this purpose select whichever operators it considers to be appropriate in the circumstances (para 8.33).*	HSC 2
29.	If the safety regulator refuses to accept a safety case or its revision it should give the reasons for that decision (para 8.33).	HSC 2
30.	*In regard to the safety case for Railtrack or any material revision, the safety regulator should give any train operator, selecting whichever it considers to be appropriate in the circumstances, the opportunity to make representations as to whether or not the safety case or revision should be accepted, and the grounds on which the representation is based (para 8.34).*	HSC 2
31.	*Railway Safety should cease to discharge the function of assessment for the purposes of the Safety Case Regulations. It should be for the safety regulator to decide to what extent, if at all, it should commission assessment from an independent body (para 8.36).*	HSC 3
32.	*A provision should be made in the Safety Case Regulations imposing a duty on Railtrack to carry out, or procure the carrying out by a suitably qualified body of, audits for the purposes presently set out in Regulation 9 of the 2000 Regulations (para 8.37).*	HSC 3
33.	*The safety regulator should review the adequacy of Railtrack's auditing, carrying out its own audits to the extent that it considers appropriate, and dealing with instances of non-compliance whenever they arise (para 8.37).*	HSE 3
34.	Regulations 12 and 13 of the 2000 Regulations should remain in effect (para 8.37).	

The safety regulator

35.	*The HSE, through the HMRI, should continue to fulfil the function of safety regulator for the railways. However, it is imperative that the HSE are provided with adequate resources in order to fulfil their role (para 9.66).*	HM Government
36.	*The HMRI should be placed under the direction of a new post, to be filled by a person of outstanding managerial ability, not necessarily with a railway background. This post should be regarded as commanding a special salary level for the purpose (para 9.66).*	HSC 2
37.	The Government should use all reasonable endeavours to ensure that standards such as Railway Group Standards are not required by the European Directive on Railway Safety in its final form to be set by the safety regulator, and that the draft Directive is modified to such extent as is necessary for that purpose (para 9.74).	HM Government
38.	The extent of passenger representation on the RIAC should be re-considered (para 9.80).	HSC 1
39.	The RIAC should be concerned with questions of safety strategy at a high level (para 9.80).	HSC

A rail industry safety body

40.	*The function of the setting of Railway Group Standards should be assumed by a new rail industry body which is independent of both Railtrack Group plc and their subsidiaries and of the safety regulator (paras 8.38, 9.46 and 10.1).*	Rail Regulator 3
41.	The body should be responsible for setting not only Railway Group Standards but also standards of the type which have ceased to be Group Standards in respect that they are concerned only with the interiors of rail vehicles (para 10.7).	Rail Regulator 3
42.	It should be considered whether in due course:	RISB

> (i) the separate existence of the SAB is unnecessary; and
>
> (ii) the RISSC should become a strategy committee of the body (para 10.8).

43.	*The body should also be responsible for the preparation of any proposed changes to the Railway Group Standards Code (para 10.8).*	Rail Regulator 3
44.	*The body should have explicit duties to set and review standards. In the performance of its duties it should be subject to the supervision of the HSE through auditing and other actions (para 10.9).*	Rail Regulator 3
45.	*The standards should be binding not only on members of the Railway Group but also on any company to which the requirement to comply currently applies, whether by virtue of a licence condition or a contractual term (para 10.10).*	Rail Regulator 3
46.	*The body should have the benefit of feedback from the auditing carried out by Railtrack and the HSE (para 10.14).*	Railtrack HSE
47.	*The body should also be responsible for the accrediting of the suppliers of products and services and the licensing of individuals, subject to the supervisory activity of the safety regulator (para 10.15).*	Rail Regulator 3
48.	The body should take an active role in steps to streamline the processes for the approval of new rail vehicles (para 10.17).	RISB
49.	*The body should be set up as a new legal entity, independent of any company in the rail industry and of any part of that industry. It should have the power and the duty to take binding decisions (para 10.21).*	Rail Regulator 3
50.	*The arrangement of the governance of the body should include provision for the representation of railway operators and of any other company to which the requirement to comply with Railway Group Standards or the additional standards referred to in Recommendation 41 applies, whether by virtue of a licence condition or a contractual term. There should also be representation of the manufacturers and suppliers of infrastructure equipment and rolling stock, and the three main rail trade unions (para 10.22).*	Rail Regulator 3
51.	*The body should have an independent chairman and a number of independent members with suitable practical experience (para 10.23).*	Rail Regulator 3
52.	*There should be a clear and easily accessible means of resolving any matter which is in dispute (para 10.23).*	Rail Regulator 3

53.	Consideration should be given to the constitution of the body by modification of Railtrack's network licence and the licences of the other railway operators (para 10.26).	Rail Regulator 3
54.	The body should be funded by means of a levy on the companies covered by the requirements referred to in Recommendation 50 (para 10.27).	Rail Regulator 3
55.	The body should also exercise a number of functions to assist the members of the rail industry to collaborate in the promotion of safety, including:	RISB 3

     (i)   establishing and managing system authorities;

     (ii)   funding and sponsoring research and development;

     (iii)   monitoring and reporting on the industry's safety performance;

     (iv)   developing the annual Railway Group Safety Plan;

     (v)   disseminating good practice; and

     (vi)   providing safety leadership (para 10.29).

56. Even if the European Directive on Railway Safety in its final form requires that responsibility for setting standards such as Railway Group Standards are to be taken over by the safety regulator, a rail industry safety body should be set up and assume the functions referred to in Recommendations 47, 48, 55 and 70 (para 10.30).

Accident investigation

57.	The responsibility of the HSE for the investigation of rail accidents should be transferred to an independent body, here referred to for convenience as the RAIB (para 9.29).	HSC 2
58.	The investigation of rail accidents and incidents of whatever nature should be brought under the overall control of the RAIB (para 11.8).	HSC 3
59.	The more serious cases should be the subject of inquiry by the RAIB. The categories of case which would fall to the RAIB to inquire into should be the subject of further study (paras 11.8 and 11.10).	HSC 3

60.	*The less serious cases should be delegated to the industry to be dealt with by formal inquiry or formal investigation. However, the RAIB should have the ability to call in any case for inquiry by itself where that appears to be appropriate (para 11.8).*	HSC 3
61.	Consideration should be given, in the longer term, to reducing the investigation of accidents or incidents at industry level to a single method (para 11.9).	RAIB
62.	*The sole objective of the investigation of accidents or incidents should be the prevention of accidents and incidents. It should not be the purpose of such investigations to apportion blame or liability (para 11.11).*	HSC 3
63.	The appointment of an independent chairman and, where appropriate, independent members for the panel of a formal inquiry, is endorsed (para 11.13).	Railway Safety RAIB
64.	Save and to the extent that there is good reason to the contrary, representatives of persons who have been affected by an accident should be allowed to attend, as observers, formal inquiries into more serious accidents. There should be a criterion for the purpose of determining for which inquiries this would be suitable (para 11.14).	Railway Safety 1
65.	Representatives of those who have been affected by an accident should be allowed to attend as observers at an RAIB inquiry into that accident (para 11.14).	RAIB
66.	Procedures, such as those followed by the AAIB, for keeping those who have been bereaved or injured fully informed of what is happening during the investigation process, are commended (para 11.15).	RAIB
67.	*The RAIB should exercise a supervisory function in regard to the working of formal inquiries and formal investigations (para 11.16).*	HSC 3
68.	The proposal of an appeal against a finding of a formal inquiry should be the subject of further study (para 11.16).	Railway Safety 1
69.	*The reports of RAIB inquiries and formal inquiries should be published, subject to the protection of the identity of persons involved (para 11.17).*	Railway Safety RAIB

70.	*The rail industry safety body should maintain a current record of:*	RISB
	(a) the recommendations of RAIB inquiries and formal inquiries;	
	(b) the responses of all the organisations to which the respective recommendations are directed; and	
	(c) the state of progress towards implementation in relation to stated timescales (paras 11.19-11.21).	
71.	*The RAIB should regularly examine the reports of formal investigations in order to determine whether there are matters of importance which should be brought to the attention of the industry (para 11.22).*	RAIB
72.	There is a need for a protocol dealing with the release of technical information and access to technical experts in investigations involving the police (para 11.29).	BTP Railtrack HSE 1
73.	The statements made by witnesses in connection with RAIB inquiries and industry inquiries and investigations should not be disclosed to the police, save by order of a judge (para 11.32).	HSC

The implementation of recommendations

| 74. | As in the case of the report on Part 1 of the Inquiry, a review of compliance with the above recommendations should be conducted on behalf of the HSC within six months of publication of this report, and further reviews should be put in hand as necessary thereafter. The HSC should publish the outcome of such reviews. | HSC |

Appendix 1
Parties and their representatives

The Inquiry

Mr Robert Owen QC, Mr Neil Garnham, Barrister, Mr Eric Brown, Barrister, Ms Susan Chan, Barrister; Mr Michael Fitzgerald, Solicitor, Mr Myles Hothersall, Solicitor, both of the Treasury Solicitor's Department, London.

Collins Passengers' Group

Mr Kenneth Hamer, Barrister; Messrs Collins, Solicitors, Watford.

Southall and Ladbroke Grove Solicitors' Group

Mr John Hendy QC, Mr Michael Ford, Barrister, Mr Rohan Pirani, Barrister; Messrs Christian Fisher, Solicitors, London, on behalf of the Southall and Ladbroke Grove Solicitors' Group.

Joint Rail Unions

Mr Jeremy McMullen QC, Mr Graham Watson, Barrister, Mr Daniel Bennett, Barrister; Messrs Pattinson and Brewer, Solicitors, London; Russell Jones and Walker, Solicitors, London; Thompsons, Solicitors, Ilford.

Association of Train Operating Companies

Mr Greg Treverton-Jones, Barrister, Mr Chris Jackson, Solicitor; Messrs Burges Salmon, Solicitors, Bristol.

Rail Users' Consultative Committees

Mr John Cartledge (lay representative).

Railtrack

Mr Roger Henderson QC, Mr Stephen Powles QC, Mr Prashant Popat, Barrister, Mr Andrew Kinnier, Barrister; Company Secretary and Solicitor to Railtrack Plc.

Health and Safety Commission and Health and Safety Executive

Mr Hugh Carlisle QC, Mr David Barr, Barrister, Ms Emma-Jane Hobbs, Barrister; Solicitor to the HSC and the HSE.

Strategic Rail Authority [1]

Mr Adrian Brunner QC, Mr George Alliott, Barrister; Solicitor to the Strategic Rail Authority.

The Rail Regulator

Mr Richard Mawrey QC, Mr James Palmer, Barrister; Director of Legal Services and Chief Legal Adviser to the Rail Regulator.

British Transport Police

Mr Richard Lissack QC, Mr Hywel Jenkins, Barrister, Mr Tom Leeper, Barrister; Solicitor to the British Railways Board.

Amey Rail

Mr Tom Custance, Solicitor Advocate of Messrs Herbert Smith & Co, Solicitors, London.

Rolling Stock Leasing Companies

Mr Philip Havers QC, Mr David Evans, Barrister; Messrs CMS Cameron McKenna, Solicitors, London.

English, Welsh and Scottish Railway

Mr Michael Mylonas, Barrister; Solicitor to English, Welsh and Scottish Railway.

Railway Industry Association

Mr Jeremy Candfield (lay representative)

Note 1

The Strategic Rail Authority were at the time of the Inquiry still in "shadow" form but for simplicity references in the report are to the Strategic Rail Authority

Appendix 2
Witnesses

Witnesses who gave oral evidence are designated "O"; witnesses who gave written evidence are designated "W".

Witnesses are, as far as possible, described either as at the time of the crash or other relevant period, as appropriate.

Abbott, H T	Managing Director, Angel Trains	O
Ackermans, F	General Manager, Safety and Regulatory Affairs, Canadian Pacific Railway	O
Alder, W A T	Principal Consultant, Entec UK	O
Atkinson, J T	Manager, Rail Safety, Land Transport Safety Authority of New Zealand	O
Bacon, J H	Director-General, HSE	O
Baker, S K	Deputy Managing Director, Northern Spirit	O
Baldry, Prof C J	Professor of Human Resource Management, University of Stirling	O
Beswick, M J	Director, Network Regulation, ORR	O
Bird, K	Chairman, c2c Rail	O
Blyth, A G	Deputy Director, Safety, Eurotunnel	W
Brabazon, P G	Formerly Principal Consultant, Entec UK	O
Brearley, S A	Controller, Safety Strategy and Planning, S&SD, Railtrack	W
Britton, R J	Secretary and Legal Adviser, CAA	W
Brown, M H	Assistant Chief Inspector of Railways, HMRI	O
Burdsall, B R	Managing Director, Midland Main Line	O
Carr, C	Technical Director, Amey Rail	O
Coleman, V P	Chief Inspector of Railways, HMRI	O
Corbett, G M N	Chief Executive, Railtrack	O
Cox, R A	Consultant Engineer	O/W
Davies, Prof J B	Professor of Psychology, University of Strathclyde; Director, CIRAS	O
Davies, Sir David	President, Royal Academy of Engineering; Chairman Designate, Railway Safety	O
Eccles, G C	Director, Stagecoach Holdings	O
Evans, Prof A W	Professor of Transport Safety, Centre for Transport Studies, University College, London	O/W
Eves, D C T	Deputy Director-General, HSE	O/W
Forster, A E	Operations and Safety Director, FGW	O
Hall, S	Expert and Author on Railway Safety, Signalling and Accidents	W
Hince, V G	Senior Assistant General Secretary, RMT	O
Holden, Major C B	Transport Safety Consultant; formerly Inspector of Railways	O
Hutton, D M	Vice-Chair, National Consumer Council	O

Kemp, Prof R J	Safety Director, Alstom Transport; Visiting Professor of Engineering, Lancaster University	O
Knapp, J	General Secretary, RMT	O
Kooger, J	Senior Consultant, DuPont Safety Resources	O
Maidment, D J	Risk Management Advisor, ERM Risk Solutions	W
McClean, R H	Production Director, GNER	O
Morris, R J	Executive Director, London South East; formerly Technical Director, Safety and Operations, SRA	O
Morton, Sir Alastair	Chairman, SRA	O
Muttram, R I	Director, S&SD, Railtrack	O
Naish, I S	Director, Investigations (Rail and Pipeline), Transportation Safety Board of Canada	O
O'Connor, P D T	Consultant Engineer	O/W
Perry, C E	Group Managing Director, AEA Technology Rail; Chairman, RISSC	O
Porter, L K	Transport Business Manager, Lloyd's Register	O
Profit, G R	Group Director, Safety Regulation, CAA	O
Rix, M D	General Secretary, ASLEF	O
Rosser, R	General Secretary, TSSA	O
Scott, I A P	Director, Safety, Health and Environment, Eurotunnel	O
Sharp, A G	Safety Risk Manager, Assurance and Safety Directorate, Railtrack	W
Smart, K P R	Chief Inspector of Air Accidents, AAIB, DETR	O
Smith, J W	Head of Regulation, Railtrack	O
Smith, Prof R A	Head of Department of Mechanical Engineering, Imperial College; Chairman, Advanced Railway Research Centre	O
Spackman, M	Consultant, NERA	W
Sylvester-Evans, R	Safety Consultant	O
Taig, A R	Managing Partner, Risk Solutions	O
Tunnicliffe, D	Formerly Chief Executive, London Transport and Chairman, LUL	O
Waite, P J	Technical Director, Entec UK	O/W
Waldram, I M	Immediate Past-President, IOSH	O
Walker, S P	Reader in Computational Mechanics, Imperial College, London	W
Walter, M H	Controller, Safety Management Systems, S&SD, Railtrack	O
Wheeler, C J	Project Manager, National Track Safety Strategy Group; Chairman, Association of On-Track Labour Suppliers; Safety Advisor, Balfour Beatty Rail Renewals	O
Wilks, D W	Infrastructure Contracts Manager, Railtrack Southern Zone	O
Williams, D J	Chief Constable, BTP	O
Winsor, T P	Rail Regulator	O
Woolfson, C A	Director, Faculty of Social Sciences Graduate School, University of Glasgow; Director, European Centre for Occupational Health, Safety and the Environment	O

Appendix 3
Principal documents

The following is a list of the principal documents which were referred to in Part 2 of the Inquiry, and are in the public domain.

Legislation

Health and Safety (Consultation with Employees) Regulations 1996

Health and Safety at Work etc Act 1974

Management of Health and Safety at Work Regulations 1999

Railway Safety (Miscellaneous Provisions) Regulations 1997

Railway Safety Regulations 1999

Railways (Safety Case) Regulations 1994

Railways (Safety Case) Regulations 2000

Railways (Safety Critical Work) Regulations 1994

Railways Act 1993

Railways and Other Transport Systems (Approval of Works, Plant and Equipment) Regulations 1994

Reporting of Injuries, Diseases and Dangerous Occurrences Regulations 1995

Safety Representative and Safety Committee Regulations 1977

Transport Act 2000

Other documents

A New Deal for Transport: Better for Everyone (DETR, 1998)

Assessment Criteria for Railway Safety Cases (HSE)

Automatic Train Protection for the Railway Network in Britain: Sir David Davies (Royal Academy of Engineering, February 2000)

Consultation Document on Transport Safety (DETR, 1999)

Council Directive 91/440/EEC of 29 July 1991 on the Development of the Community's Railways

Ensuring Safety on Britain's Railways (HSC, 1993)

Ensuring that Railtrack Maintain and Renew the Railway Network: Report by the Comptroller and Auditor General (HMSO, April 2000)

Internal Inquiry Report: Events Leading up to the Ladbroke Grove Rail Accident (HSE, April 2000)

Joint Inquiry into Train Protection Systems: Professor John Uff QC and the Rt Hon Lord Cullen PC (HSC, 2001)

Maintaining a Safe Railway Infrastructure: a Report on Railtrack's Management System for Contractors (HSE, 1996)

Major Incident Response and Investigation Policy and Procedures (HSE, June 1999)

New Opportunities for the Railways: The Privatisation of British Rail (HMSO, Cmnd 2012, July 1992)

Notice of Modification to Condition 3 of Railtrack's Network Licence (ORR, October 2000)

Proposals for The Railways (Safety Case) Regulations 2000 (HSE, February 2000)

Railtrack's Safety and Standards Directorate: Review of Main Functions and their Locations (DETR, February 2000) (the "Rowlands Report")

Railway Group Standards Code (Railtrack, June 1998)

Railway Safety Cases: Guidance on Railways (Safety Case) Regulations 1994 (HSE, 1994)

Railway Safety Critical Work: Approved Code of Practice (HSE, 1994)

Railway Safety in Japan: Mission Report (DTI/Advanced Railway Research Centre, 2000)

Railway Safety Principles and Guidance (HSE, 1996)

Railway Safety: Environment, Transport and Regional Affairs Committee, First Report (HMSO, November 1998)

Railways (Safety Case) Regulations 2000: Draft Guidance (HSE, October 2000)

Regulating Higher Hazards: Exploring the Issues (HSE, 2000)

Report on some General Issues Arising from the Internal Inquiry into Events Leading up to the Ladbroke Grove Rail Accident (HSE, April 2000) (*the "General Issues Report"*)

Report on the Inspection carried out by HM Railway Inspectorate during 1998/99 of the Management Systems in the Railway Industry covering Signals Passed at Danger (HSE, September 1999)

Review of Arrangements for Standard Setting and Application on the Main Railway Network (HSE, September 1999) (*the "Tansley Report"*)

Safety Regulations and Standards for European Railways (National Economic Research Associates (NERA), February 2000)

Southall Rail Accident Inquiry Report: Professor John Uff QC (HSC, 2000)

Successful Health and Safety Management (HSE, 1997)

The Management of Safety in Railtrack (HSE, February 2000)

Transport Safety Review: Consultation Response and Consideration of Issues (DETR, June 2000)

Appendix 4
The relevant accidents

Newton Junction (21 July 1991)

1. This accident involved a head-on collision between two suburban electric passenger trains on a short stretch of single track west of Newton station, near Glasgow. The Balloch-Motherwell train was approaching Newton station at the time, while the Newton-Glasgow train was leaving it. Both drivers and two passengers were killed, and 22 passengers were injured.

2. The cause of the accident was found to be the passing at danger of the platform starting signal (M145) at Newton station by the Newton-Glasgow train. The Balloch-Motherwell train was signalled at the time to proceed into Newton station.

3. Key features of the accident included the fact that several trains were running late at the time, with four train movements scheduled to occur within a nine minute period. The single lead junction into the station on which the accident occurred had been only recently installed, as part of a major redevelopment of track and signalling. Signal M145 had been passed at danger by a Driver-Only Operation (DOO) train just a month beforehand. The accident was found to be similar to two previous incidents (Bellgrove and Hyde).

4. The HSE inquiry into the accident made recommendations relating, amongst other things, to:

    *   the reinstatement of double lines at the junction concerned;
    *   the management and commissioning of track layout and signal installation projects;
    *   risk analysis for proposed schemes involving single-track working;
    *   the training of drivers relating to new signalling schemes;
    *   track-to-train radio communications; and
    *   ATP.

Watford South (8 August 1996)

5. In this accident a North London Railways passenger train and an empty coaching stock train collided some 700m south of Watford Junction Station. The passenger train had passed a signal at danger. One passenger was killed, and 69 passengers and four train crew members were injured.

6. The passenger train was travelling north from Euston to Milton Keynes on the down slow line. The driver had not reacted correctly to two signals set at caution by slowing down and preparing to stop. When he saw the following signal at red he applied the brakes but the train (which had been travelling at 68 mph) stopped 203m past the signal, across a junction linking the down slow line to the up fast line. The

southbound coaching stock train, passing over this junction, was unable to avoid a collision with the stationary passenger train.

7.   The HSE inquiry concluded that:

- the collision would have been avoided if ATP had been fitted and in operation;
- the wording of a Railway Signalling Standard was imprecise. This led to a speed restriction sign being placed in an inappropriate position, which gave confusing information to the train driver; and
- the signal that was passed at danger had a shorter than normal overlap.

8.   Most of the inquiry recommendations required Railtrack to take on a pro-active and co-ordinating role so that they could be satisfied that risk was being properly controlled on their infrastructure by the train operators and others.

Bexley (4 February 1997)

9.   A freight train consisting of two locomotives and 19 wagons, and carrying spoil from railway track renewal work, derailed on a bridge just after passing through Bexley station, Kent. Although the two locomotives and the first 11 wagons remained on the track the twelfth was partially derailed, and the following seven wagons left the track completely, causing extensive damage. Four members of the public were injured.

10.   The HSE inquiry found that the primary cause of the derailment was lateral movement of the track on the longitudinal wheel timbers on the bridge, where significant deterioration of the track support and its fastenings had occurred. Two other factors contributed to this primary cause: the twelfth wagon was overloaded, and the train was travelling above the speed limit for freight trains on the route. The section of track concerned had been identified on a number of previous occasions to be in urgent need of repair.

11.   A number of underlying causes, reflected in the HSE's recommendations, were also identified:

- management failure of the IMC;
- failure of the TRC to ensure safe loading of the wagons;
- failure by Railtrack to monitor the performance of the IMC;
- inadequate training of the train driver; and
- inadequate arrangements for the inspection, maintenance and calibration of the locomotive speedometers.

Newton Abbott (6 March 1997)

12.   A Paddington-Penzance HST, travelling at around 60 mph, derailed as it approached Newton Abbott station. Eight passengers were taken to hospital.

13.   The derailment was caused by an axle failure on the second coach of the train. The failure was caused, in turn, by a crack in the axle journal, initiated by fretting fatigue between the axle journal and the bearing. The indications were that cracks were present in the axle journal before the wheelset was last overhauled in January 1997. They were not detected by the testing methods used, although those methods were subsequently confirmed as able to detect cracks if correctly applied.

14.   Following the derailment the train crew failed to communicate with each other in accordance with the Rule Book, and to prevent passengers from leaving the train before the site had been protected. Evacuation procedures were not properly understood by the train crew.

15.   The recommendations of the Railtrack Great Western Zone formal inquiry report related to:

   - the audit and validation by TOCs of their suppliers;
   - the testing regime;
   - testing equipment; and
   - the need to revisit the recommendations relating to emergency procedures which followed the inquiry into the train fire at Maidenhead (8 September 1995).

In addition it was recommended that the inquiry report should be forwarded to those implementing the recommendations of the inquiry into the derailment at Rickerscote (8 March 1996), which was also caused by axle failure due to metal fatigue.

Southall (19 September 1997)

16.   The accident took place at Southall East Junction, West London, when a HST from Swansea to Paddington collided with a freight train crossing the up and down main lines to Southall Yard. The collision resulted in the death of seven passengers on the HST and many injuries.

17.   The decision to route the freight train across the up and down main lines was in accordance with the rules. The junction was protected at the time of the collision by three signals on the up main line. The HST driver (who had taken over the train at Cardiff) failed to heed the first two of these. He braked on seeing the third, at red, but the trains were still travelling at a relative speed of above 80 mph when the collision occurred.

18.   The HST had travelled from Swansea with the AWS isolated. A fault had been reported on the previous day but testing at the maintenance depot overnight did not reveal any fault and the train was passed for service. The AWS failed again early in the morning of 19 September at Paddington Station, where it was isolated by the

driver at the time. The same driver reported the problem, but not, as the Rules required, to the Signalman. Fitters attended the train at Swansea but did not attempt to repair the AWS. No action was taken to withdraw the train from service or to turn it so that the leading power car had an operational AWS. The ATP fitted to the train was not switched on, because neither driver on the Swansea/Paddington journey was currently qualified to drive with ATP.

19.   The primary cause of the accident was the driver's failure to respond to the two warning signals. Other causes were the failure of the train operator's maintenance system to identify and repair the AWS fault; the train operator's failure to react to isolation of the AWS; the failure of Railtrack to put in place rules to prevent normal running of a HST with AWS isolated; and the train operator's failure to manage the ATP Pilot Scheme such that the ATP equipment was switched on.

20.   Professor John Uff QC, who chaired the public inquiry into the crash, made 93 recommendations under the following headings:

- driver training;
- operating rules;
- fault reporting;
- fleet maintenance;
- infrastructure maintenance;
- regulation;
- vehicle design;
- research and development;
- ATP;
- general safety issues;
- accident investigations and inquiries; and
- post-accident procedures.

Norton Junction (23 February 1998)

21.   This incident involved the wrong-side failure of the AWS ramp at a permanent speed restriction at Norton Junction, on the Evesham-Worcester line. Two trains on 23 February reported that there was no AWS indication at the ramp. A fault team arrived on site and declared the fault restored. On 24 February two further trains reported no AWS indication. Again a fault team attended and declared the fault restored. A team was dispatched a third time on 25 February after a further report, with the same result. On 26 February a team arrived on site without being called, found a cable fault, and rectified it. No further faults occurred.

22.   The Railtrack formal inquiry into the incident found that the immediate cause of the problem was worn insulation on a cable which had caused intermittent faults with the AWS equipment. The underlying cause was the failure of the local response team to identify the fault as a wrong-side failure and then test to the required standard. Furthermore the fault control centre had failed to identify on each occasion that a wrong-side failure had occurred.

Ladbroke Grove (5 October 1999)

23. The report for Part 1 of the Inquiry gave a detailed account of this crash, in which a three car Turbo train bound for Bedwyn passed a signal at red at Ladbroke Grove, about two miles to the west of Paddington. The Turbo was carried towards the path of a HST approaching Paddington. The two trains collided at a combined speed of around 130 mph. 31 people were killed, including both drivers. Over 400 other persons suffered injuries, some of them of a critical nature.

24. It was concluded that the poor sighting of the signal, allied to the effect of bright sunlight at a low angle behind the driver, probably led him to believe that he had a proceed aspect. The unusual configuration of the signal not only impaired the initial sighting of its red aspect but also might well have misled an inexperienced driver such as the driver of the Turbo. He had only recently qualified as a driver and there were significant shortcomings in his training.

25. A signaller had put back a signal to Danger in a vain attempt to halt the HST. An emergency stop message was sent to the Turbo, but it was not possible to determine whether it was received before the crash. For a period of time after being alerted to the SPAD the signaller did not take action as he was expecting the driver of the Turbo to stop within the overlap. Serious deficiencies in the running of the signalling control centre were revealed. The Inquiry also found wider deficiencies in Railtrack's management of the Zone. These included a lack of adequate consideration of the difficulties faced by drivers and failures to convene signal sighting committees, carry out risk assessment, respond to the recommendations of inquiries and pursue improvements effectively.

26. The recommendations made in the Part 1 report covered a number of subjects including:

- track and signalling changes;
- signalling in the Paddington area;
- the implementation of recommendations of rail industry inquiries;
- driver management and training;
- signal sighting;
- SPAD investigation;
- signallers' instructions and working conditions;
- control centre equipment and radio communications;
- crashworthiness and fire mitigation; and
- passenger protection, evacuation and escape.

Appendix 5
The models proposed by parties to the Inquiry

Introduction

1. This appendix provides an outline of the models which were proposed by parties to the Inquiry. It is based on parties' statements of case, as amended in the course of Part 2, and their closing submissions.

2. The models proposed were, broadly speaking, a new safety regulator or a new standards body for the industry. As can be seen, the amount of detail which was proffered by the parties varied considerably. No model was proposed by Amey Rail or EWS.

3. All the parties, apart from the HSE, Amey Rail and the Joint Rail Unions, proposed that the function of railway accident investigation should be transferred from the HMRI to a new rail accident investigation body. The HSE were opposed to this. Amey Rail offered no view. The Joint Rail Unions stated that they would be content with a separate division within the HMRI taking charge of this.

ATOC

4. ATOC proposed a new NRSA, which would supersede the HMRI and Railway Safety. The NRSA would be separate from the HSE and operate autonomously under an agency agreement, reporting to the Department of Environment, Transport and the Regions (DETR). There would be a memorandum of understanding between the NRSA and the HSE which would define the boundary between them as being the ticket barrier, and would provide for the NRSA assuming jurisdiction where there was an overlap between the two bodies, as in the case of track workers.

5. The functions of the NRSA would include:

    *   standard setting;
    *   acceptance of RSCs;
    *   acceptance of vehicles;
    *   accreditation (a system under which companies supplying equipment and services to the rail industry would be directly accountable);
    *   auditing;
    *   enforcement;
    *   improvement and prohibition notices;
    *   prosecution;
    *   data capture – trends;
    *   exchange of information;
    *   promulgation of best practice and lessons;
    *   establishment and management of system authorities; and

- safety research – funding and sponsoring.

6. A large part of the funding for the NRSA would be provided by the Government, but otherwise it would be provided by members of the rail industry on a pro rata basis. Licensing and approvals could be at least partly self-financing, e.g. a fixed (and regulated) fee could be set by the NRSA, after consultation with the industry, and be subject to review. The S&SD and the HMRI had skilled personnel who could fill some of the key positions in the NRSA, but an infusion of new blood and new approaches were essential. To encourage recruitment civil service pay scales would be relaxed, and some accommodation offered to middle-aged skilled personnel in order to attract them to undertake work for a little less than they would earn in the industry.

7. It should be noted that:

    (i)   RISSC and the standards subject committees would be adapted for the task of drafting standards for submission to the NRSA through a "Rail Standards Setting Executive". This body would also act as a forum for co-ordinating industry groups;

    (ii) the results and contents of audits would be made available to all relevant stakeholders;

    (iii) the NRSA would be responsible for co-ordination, strategic prioritisation and follow-up of recommendations. Those which were directed towards the NRSA would be checked by the DETR.

8. A new consultation body, the Rail Safety Consultative Body, would replace the SAB and the RIAC.

ROSCOs

9. The ROSCOs supported the model of the NRSA which had been proposed by ATOC. They added that the NRSA would:

    (i)   be funded by a levy on all holders of safety cases and accredited suppliers, which would be based on the turnover of the industry;

    (ii) be advised by a committee of senior members of the industry and independent persons with extensive knowledge of it; and

    (iii) have the power, in accordance with clear and consistent criteria, to fine and to suspend, or revoke the acceptance of, safety cases and accreditation.

The bereaved and injured represented by the Southall and Ladbroke Grove Solicitors' Group

10. This group also proposed a NRSA, which would be similar to that proposed by ATOC but would have a number of additional features. They would be a public body established by statute. Their board members would be appointed by the Secretary of State, in consultation with the HSC, to represent the public interest in a safe railway system. They would be persons of relevant experience and railway commitment, mainly independent-minded persons from a railway background. They would not be appointed as representatives of particular parts of the industry. However, bodies such as the IRSE, the CAA and the HSC would be invited to make nominations. Their appointments would be for a fixed term and terminable only on specified grounds.

11. The core of the staff of the NRSA would be drawn from the HMRI and Railway Safety, but their numbers would need to be increased. Remuneration would be comparable with that in the industry. The funding of the NRSA would comprise the funding presently provided for the HMRI and Railway Safety, along with a levy on the industry and modest charges above the cost of certification.

12. A body (referred to as a pan-industry body) would also be set up by the industry parties to represent their interests and foster co-operation and collaboration. It might be constituted by members of the Railway Group along with other relevant parties, such as the ROSCOs and manufacturers. Most of its work would be done by committees representing only those parties which had interest in the particular subject matter. It would be involved in the setting of strategic safety targets. Many of the functions of the NRSA would depend on drafts and representations by this body (or other industry groups). There would still be a role for ATOC, RISSC and the subject committees.

13. The functions of the NRSA would include:

   - formulation of safety strategy;
   - inspection and enforcement, including improvement and prohibition notices and prosecution;
   - certification, including the power to suspend or withdraw certificates;
   - acceptance of safety cases;
   - audits;
   - approval of Group Standards;
   - approval of equipment;
   - research and development;
   - system authorities;
   - tracking of inquiry recommendations; and
   - "residual functions".

14.   It should be noted that:

(i)   the NRSA would be responsible for scrutinising the safety goals identified by industry, and for setting safety strategy where the industry had failed to do so;

(ii)   the NRSA would have power to impose appropriately graded sanctions; and power, through the Rail Regulator, to grant incentives and impose penalties. This should be subject to a right of appeal to an internal panel;

(iii) the standards to be sanctioned by the NRSA would be drafted by the pan-industry body which would attempt to agree them prior to their submission. The NRSA would have power to propose and (after consultation) impose amendments to the drafts. They would have power themselves to draft and impose standards if the industry was unable to reach agreement or delayed in doing so. Whether there should be an appeal process was "an open question";

(iv)   in regard to system authorities, the NRSA would have an overarching power. They would have a particular role when the industry had difficulties, especially when agreement could not be reached as to the way in which costs should be borne;

(v)   as regards research and development, the NRSA would have the power to raise funds by way of levy on the industry (or parts of it) and to seek grants from the SRA; and

(vi)   the NRSA would collate, monitor and, when necessary, pursue recommendations of inquiries and investigations. They would hold and publish a cumulative list of recommendations, the responses to them (including reasons for non-acceptance) and progress in implementation. They would investigate non-acceptance and should have power to require implementation.

15.   The NRSA would be under a duty to consult with all persons likely to be affected by their decisions and activities, including the pan-industry body and other industry groups. A key internal group is the workforce.

16.   The NRSA would observe "transparency". For example, members of the public would be permitted to observe their full meetings. All their decisions and reports would be published. Once again a key interest group was the workforce.

17.   The NRSA would have the right, on safety grounds, to veto the choice of a franchisee and to revoke a franchise. On similar grounds they would have the power to veto a proposal by the Rail Regulator.

The SRA

18. The SRA proposed a new separate rail safety authority, on the model of the SRG in the CAA. This authority would also perform the function of licensing of personnel, equipment etc.

19. The authority would be formed out of Railway Safety and a "revitalised HMRI". Its staff would be paid salaries at industry levels.

20. Its functions would include:

     • the adoption of safety strategy;
     • standard setting and compliance;
     • incorporation of system authorities; and
     • acting as a catalyst for co-ordination and direction of research and development.

Rail Users' Committees

21. The Rail Users' Committees proposed an autonomous rail safety executive. This would be concerned with safety "across the board", but not including occupational health and safety. It would replace the HMRI and would have the same responsibilities, apart from accident investigation.

22. The rail safety executive would not be obliged to compete for resources in the way that other divisions of the HSE had to do. It would recruit in the market for the best available personnel at appropriate rates. It would not be restricted to the goal-setting style of regulation if and where a more prescriptive approach was necessary to meet the expectations of society and to overcome resistance to its guidance within the industry. It would be able to commission advice and services from external sources.

23. The rail safety executive would foster close consultative links at the policy-making level with all relevant parties within the industry. It would take over administrative responsibility for the RIAC, and perhaps the National Safety Task Force which was sponsored by the DETR. These would be upgraded to become a standing Railway Safety Policy Commission. The commission would meet in public to consider the whole range of rail safety issues at a strategic level, and would provide advice to the safety executive, the Government, and the industry. Its composition would encompass and reflect all the interests affected by the remit of the safety executive, including those of the travelling public.

24. The rail industry as a whole (including ROSCOs and IMCs) would take joint ownership of Railway Safety and fund them collectively, in proportion to the turnover of individual companies. The composition of its board would be such as to ensure that no single company or sectional interest could overrule the rest. It would function and be seen to function, as an independent and objective standards authority. It would answer to the safety regulator and to the public at large for its policies, priorities and performance.

Railtrack

25. Railtrack proposed a railway safety regulation agency (referred to as the Statutory Safety Regulator) which would be established under statute and independent of the HSE. It would be staffed by personnel from the HMRI and the industry with salaries at industry level. For safety professionals it would provide an opportunity for career development. It would be a new start successor to the functions previously exercised by the HMRI.

26. The functions of the Statutory Safety Regulator would include:

    * appeals in regard to the safety aspect of Railway Group Standards;
    * promoting regulation by statutory instrument, including a system authority process;
    * high level principles and guidance ("regulatory standards");
    * approval of the Railway Group Standards Code;
    * approval of safety cases;
    * approval of new works, plant and equipment;
    * audit of Railway Safety and ad hoc;
    * monitoring;
    * enforcement;
    * prosecution under the 1974 Act; and
    * consultation.

27. The "regulatory standards" would be primarily goal-setting but would be more specific than the HSE guidance and principles. They would be developed in consultation with the industry and encourage innovations leading to safety improvements.

28. In the case of an appeal in regard to a Group Standard, it would be presumed that it related to safety and accordingly the appeal would lie to the Statutory Safety Regulator. If it was shown to relate to a non-safety matter, the appeal would lie to the Rail Regulator but respecting the former's views.

29. The function of approving safety cases would be supported by Railway Safety and the industry.

Joint Rail Unions

30. Under the proposal of the Joint Rail Unions, Railway Safety would be replaced by an independent body owned and funded by the industry, with a trust-based constitution. It would be governed by a board of management. It would perform a recommendation function in regard to safety cases. It would have no responsibility for regulation. It would offer competitive terms and conditions in order to recruit expert and qualified staff.

31. Railtrack Line would be expected to expand its Assurance and Safety Directorate.

Collins Passengers' Group

32. The Collins Passengers' Group proposed a new Rail Standards Authority. It would be designated by statute as the authority on technical and competence issues. It would be representative of the industry as a whole, but independent of its members. It would be funded by the industry, probably through a compulsory levy.

33. It would incorporate the technical division of the HMRI (Division RI1) and the Industry Safety Liaison and Technical Services Departments of Railway Safety.

34. Its functions would include:

    - setting, and ensuring compliance with, Railway Group Standards;
    - giving the lead in matters of operational safety by operators, contractors and sub-contractors;
    - acting as an overarching body to undertake research and development;
    - acting as a system authority;
    - seeing that projects across the industry are carried through to completion within sensible timescales;
    - resolving conflicts over responsibility for technical safety arising from fragmentation;
    - accrediting contractors and sub-contractors; and
    - providing a focus for direct accountability for companies supplying equipment and services to the industry.

35. The Rail Standards Authority would be separate from the safety regulator, whichever body that might be. It was envisaged that the Safety Management Systems and Safety Strategy and Planning Departments of Railway Safety would pass to the HSE, who would retain their existing functions, apart from accident investigation. It would also be responsible for higher level inspection of the Rail Standards Authority, i.e. examining its targets and achievements rather than policing details, and for dealing with any necessary appeals.

36. In addition there would be a Rail Safety Advisory Commission or a high level advisory committee within the HSE. It would be representative of a wide range of stakeholders, including passengers, trade unions, employers and suppliers. It would be "strongly representative of passenger interests", and would "engage passengers more openly and positively in the setting of safety goals and the scrutiny of regulatory processes".

The Rail Regulator

37. The Rail Regulator proposed a new Rail Standards Authority which would be separate from, and supervised by, the safety regulator. It would have an independent chairman. It would provide executive leadership.

38.   Its functions would include:

- setting standards throughout the industry;
- certifying personnel, equipment and procedures, and certifying schemes for assessing the competence of personnel and suppliers, and the safety and design of equipment.

39.   The Rail Standards Authority's concern with standards would include maintaining and constantly updating them. The standards would extend to safety, procedures, interfaces and other operational matters. In regard to non-safety matters, an appeal would lie to the Rail Regulator.

40.   It would fall to the safety regulator to ensure that Group Standards met current requirements of safe legislation and were in accordance with good practice. It would exercise a supervisory function, and would have a duty to prompt the Rail Standards Authority to action where this was required. The safety regulator would be the verification agency for safety cases.

41.   The Rail Regulator was neutral as to whether the safety regulator should remain the HSE or whether that function should be devolved to a new safety authority.

The HSE

42.   The HSE proposed an industry standards setting body, which would be wholly separate from Railtrack and would be representative of all sides of industry, including the trade unions. It would be non-statutory.

43.   The purpose of this body would be to set standards of best practice. They would cover all aspects, rather than merely those relating to Railtrack's control of risk. An emphasis on goal-setting standards would be expected, but some prescription would be appropriate. It would also deal with the accreditation and certification and the quality and competence of industry suppliers, contract labour and individual workers.

44.   Given that Railtrack would still have responsibility in regard to the importation of risk on to the network, many of the functions of Railway Safety would continue to be essential as part of Railtrack's Assurance and Safety Directorate.

45.   The HSE supported the creation of a rail industry body for research and development.

Appendix 6
Safety regulation and accident investigation in aviation

Introduction

1. This appendix provides an outline of the systems of safety regulation and accident investigation in the civil aviation sector. It is based on the evidence given to the Inquiry by:

    - Mr G R Profit, Group Director of Safety Regulation in the CAA;
    - Mr K P R Smart, Chief Inspector of Air Accidents in the AAIB; and
    - Mr R J Britton, Secretary and Legal Adviser in the CAA

    and on "the Regulatory Regime for the Railways" by Mr E C Brown, advocate and barrister.

Safety regulation

2. The Safety Regulation Group (SRG) is the safety regulator for aviation in the United Kingdom. It is part of the CAA who were established in 1972 and are independent of Government. The CAA are also the economic regulator, and until recently provided air traffic control services through a subsidiary. The Group Director of the SRG is a member of the board of the CAA.

3. The SRG is funded entirely by the industry through a statutory charging scheme. Costs are allocated throughout the industry on an equitable basis and not in proportion to services rendered.

4. The staff of the SRG are not civil servants. The SRG offers salaries which are considered to be equivalent to mid-industry levels.

5. The SRG regulates safety only with respect to the aircraft, its passengers and its cargo. It does not have responsibility for, or seek to regulate on, health and safety of those who work in the aviation industry. The boundary between the responsibilities of the SRG and those of the HSE is the subject of a memorandum of understanding between them.

6. The principal activities of the SRG are organised into three divisions which deal respectively with the separate elements in the aviation industry, namely:

    (i)   the designers and manufacturers;

    (ii) the operators; and

    (iii) the support infrastructure, i.e. air traffic control and airports.

7. The SRG is responsible for safety policy and for regulatory requirements. The regulations are a mixture of prescriptive and goal-setting requirements. Examples of the former are those relating to the limits of flying time, and the air worthiness of aircraft. Most of the regulations are set at European level through the Joint Aviation Authorities. The regulations are being aligned with the requirements of the FAA in the United States of America. 20% of the regulations are still set at national level.

8. A large number of activities require a certificate of approval from the SRG. For example an operator requires to obtain an air operator's certificate. The SRG also administers a regime for the licensing of individuals, namely pilots, maintenance engineers and air traffic controllers. This depends upon their satisfying standards which have been approved by the SRG. Thus pilot training relates under a form of approval and licensing and to an approved syllabus. Designers and manufacturers have to operate in accordance with a common European standard. Each organisation is approved by the SRG as meeting that standard. Their products are also certified under it.

9. If the maintenance of aircraft is put out to contract, the airline remains the duty holder in regard to that activity unless the contractor is approved by the SRG, in which case it is the duty holder for that purpose.

10. All activities requiring a certificate of approval from the SRG are monitored by it. The SRG's inspectors fly with airlines, and maintenance activities are subject to oversight by it.

11. The SRG has available to it a wide and graduated range of intervention and enforcement measures. It does not have the power to impose fines, but it issues verbal and written warnings. If these are not complied with, some form of sanction will follow. It may cancel the registration of an aircraft, and revoke, suspend or vary a certificate, licence, approval, validation or rating. It is required to serve notice of its proposed course of action together with the reasons for it. The party on whom the notice is served may request that the CAA themselves should decide the matter. Such "appeals" are heard by non-executive members of the board of the CAA. The CAA are also empowered to take provisional action pending the inquiry into or consideration of the case. An appeal lies to a court of law against a decision that a person is not fit to hold a licence or to act as a member of a flight crew, an aircraft maintenance engineer or an air traffic controller. If the court reverses the decision, the CAA have to give effect to that determination.

Accident investigation

12. The investigation of accidents and serious incidents is the responsibility of the AAIB, which, like the MAIB, is part of the DTLR. The AAIB is totally independent of the SRG. Air accident investigation has always been separate from safety regulation. The AAIB has its origin in the accident investigation branch of the Royal Flying Corps. It was transferred to the Ministry of Civil Aviation in 1946, became part of the then Department of Transport in 1983 and acquired its present name in 1987.

13.   The AAIB is accordingly funded by Government. The salaries paid to its staff are lower than those which they could earn in the aviation industry. The AAIB uses experts to support its investigations to the extent that this is necessary.

14.   The activities of the AAIB are governed by The Civil Aviation (Investigation of Air Accidents and Incidents) Regulations 1996 which implemented in the United Kingdom the Council Directive 94/56/EC. It may be noted that, reflecting Annex 13 to the Chicago Convention on International Civil Aviation 1944, the Directive provides that the investigation body or entity

"Shall be functionally independent in particular of the national aviation authorities responsible for air worthiness, certification, flight operation, maintenance, licensing, air traffic control or airport operation and, in general, of any other party whose interests could conflict with the task entrusted to the investigating body or entity".

15.   Regulation 4 of the 1996 Regulations states:

"The sole objective of the investigation of an accident or incident under these Regulations shall be the prevention of accidents and incidents. It shall not be the purpose of such an investigation to apportion blame or liability".

16.   Regulation 2 of the 1996 Regulations defines an "accident" as, put shortly, an occurrence associated with the operation of an aircraft which takes place between the time any person boards the aircraft with the intention of flight until such time as all persons have disembarked, in which:

(a)   a person suffers a fatal or serious injury in consequence of certain specified circumstances relative to the aircraft;

(b)   the aircraft sustains damage or structural failure; or

(c)   the aircraft is missing or completely inaccessible.

An "incident" is an occurrence, other than an accident, associated with the operation of an aircraft which affects or would affect the safety of operation. A "serious incident" is an incident involving circumstances indicating that an accident nearly occurred.

17.   The Chief Inspector of Air Accidents is under a duty to carry out, or cause an inspector to carry out, an investigation into accidents and serious incidents which occur in certain specified circumstances. These include accidents and serious incidents which occur in or over the United Kingdom and accidents which occur abroad to aircraft registered in the United Kingdom when an investigation is not carried out by another state. The Chief Inspector has a discretion whether or not to investigate or instruct an investigation into an incident, other than a serious incident. An air accident investigation will proceed irrespective of any criminal investigation or proceedings or any other form of inquiry which may be ongoing.

18. It may be noted that incidents with a potential to affect safety have to be notified to the AAIB and to be reported to the CAA under the Mandatory Occurrence Reporting Scheme. They are investigated by the SRG. The Chief Inspector of Air Accidents sees the summaries of these investigations and can commission an investigation by the AAIB at any time. The SRG will in any event draw to the attention of the AAIB anything which is considered should be investigated by the AAIB.

19. Investigation into air accidents or incidents take place in private.

20. The AAIB aims to ensure that all parties are aware of what is taking place in the investigation. Care is taken to see that manufacturers are not able to influence its actions.

21. On completion of an investigation the investigating inspector is required to prepare a report, a copy of which is submitted to the Secretary of State. It is required to contain, where appropriate, relevant safety recommendations. The regulations state that a recommendation shall in no case create a presumption of blame or liability for an accident or incident. The report must protect the anonymity of the persons involved in the incident. It must be circulated to the parties likely to benefit from its findings with respect to safety. The reports have to be made public in the shortest possible time, if possible within 12 months of the date of the accident or incident, in such manner as the Chief Inspector thinks fit. If, in the opinion of the investigating officer, publication of a report is likely to adversely affect the reputation of any person (or a deceased person), notice must be given to that person (or to the representative of the deceased) by the investigating officer. The notice must include particulars of any proposed analysis of facts and conclusions as to the cause or causes of the accident or incident which may affect the person on whom the notice is served. In practice, the investigating officer sends the whole report to the person. The inspector is required to consider any representations made prior to publication of the report. In practice, where substantive representations are made the person (or his representative) is invited to a meeting with the investigating officer at which matters are discussed.

22. If a report contains safety recommendations the report and the recommendations have to be communicated to the relevant undertakings or national aviation authorities. Any undertaking or authority to which a safety recommendation is communicated must take that recommendation into consideration and, where appropriate, act upon it. It must also send to the Secretary of State full details of the measures, if any, it has taken or proposes to take to implement the recommendation and, in a case where it proposes to implement measures, the timetable for securing that implementation. If the undertaking has decided not to implement a recommendation, a full explanation has to be given to the Secretary of State as to why that recommendation is not to be the subject of implementation measures.

23. The CAA are required to publicise their response to recommendations made by the AAIB and directed to them. This is done in the form of a progress report which is published annually by the CAA. The progress report also contains the responses of all other undertakings and authorities to which AAIB recommendations have been directed.

Appendix 7
Joint statement of experts on risk management

Introduction

1.   This appendix is based on the joint statement to the Inquiry by experts on the use and application of risk assessment in the rail industry. All the experts, who are named in para 2.23, participated in the meeting in their individual capacity.

General use of risk assessment and risk management

2.   Risk assessment and risk management are distinct. A description of what these comprise, in the context of safety issues, is given in Annex A. It is based on the Royal Society's 1992 Report on Risk Analysis, Perception and Management. The term "risk evaluation" as a separate entity was considered confusing. It should be viewed as an integral part of the risk management process. It was therefore omitted as a term, but included in the risk management description as "appraisal of the assessed risks".

3.   Risk assessment provides a tool to better inform managers of hazard and risk issues. It should assist clear thinking. By way of explanation, one of the principal benefits of risk assessment is that it examines the causes and consequences of undesired events, and may provide pointers as to how they could be avoided.

4.   The carrying out of a risk assessment does not remove from managers the responsibility for safety decisions, but it represents an important input into the decision-making process.

5.   The scope of every risk assessment should be stated explicitly. By way of explanation, this is considered to be a fundamental step that is often overlooked. In the rail industry the scope of an assessment may cross many company and system boundaries. Accordingly, the scope should specify, *inter alia*, what is included and excluded in the assessment, with particular reference to the nature and sources of hazard, the systems and people who may create the undesired events and contribute to the risks, and the systems and people who may be affected.

6.   Risk assessment should be a collaborative effort. It should include input from all relevant persons and groups, commensurate with its scope. By way of clarification, the risk assessment process should include input from those who create or contribute to the risks. Risk assessment may benefit from, but does not necessarily require, input from those who are affected by the risks. However, the views of the latter should be taken into account during the risk management process.

7.   All assumptions made in a risk assessment should be transparent.

8.   The effort put into a risk assessment, and its level of detail, should be proportionate to the perceived risk and commensurate with the scope.

9.   Both qualitative and quantitative risk assessment methods have their role. Irrespective of the methods an assessment should achieve an appropriate understanding of the hazards, failure mechanisms and interactions between systems (including human error), together with the potential nature and severity of the effects.

10.   A risk assessment should make the major uncertainties explicit. It should include a sensitivity analysis of the effects of key assumptions and dependencies.

11.   A quantitative risk assessment should be neither optimistic nor pessimistic. It should present as accurate a picture of the risks and uncertainties as is possible.

12.   Risk assessments should be checked for consistency with relevant empirical data, internal consistency, quality and reasonableness.

13.   If any additional weighting is given to some risks, for example, due to the public aversion to the nature or magnitude of a particular risk, this should be applied as part of the risk management process rather than the risk assessment process. By way of explanation, risk assessment should focus on analytical issues while "societal concerns" should be addressed during the risk management phase. This avoids the potential for "double counting". In the opinion of some experts but not all, considering "societal concerns" within risk assessment is a problem with the HSE discussion document entitled "Reducing Risks, Protecting People".

14.   The use of stated "good practice" as a substitute for undertaking a risk assessment should be used with caution. By way of explanation, "good practice" may not be as "good" as is thought. It should not be used as a substitute for clear thinking. The technique of risk assessment is powerful in challenging so-called accepted standards and situations.

15.   A poor quality risk assessment is a liability as it may well lead to poor decisions and misallocation of resources.

16.   In summary, a good risk assessment should ensure the use of a systematic and disciplined process and approach; be conducted by competent people; be proportionate to risks involved; with the clear presentation of findings describing the risks and highlighting both the conclusions and limitations of the assumptions and data; so as to inform risk management decisions.

17.   There is a convergence of accepted good practice in the <u>principles</u> of risk assessment described in the more recent British & International Standards. Mr O'Connor dissented from this view. His view is set out in Annex B.

18.   The 1999 edition of the Management of Health and Safety at Work Regulations Approved Code of Practice provides useful high-level guidance on the principles of risk assessment and what constitutes a "suitable and sufficient" risk assessment.

19.   More detailed guidance should be prepared by the safety regulator as to what constitutes a "suitable and sufficient" risk assessment for the rail industry.

Specific use of risk assessment in the rail industry

20. For those experts who are in a position to know, the quality of the risk assessments used in the rail industry has been variable. There are good and poor examples. By way of clarification, risk assessments have suffered from one or more of the following deficiencies: being superficial, too restrictive or poorly scoped, too generic, overly mechanistic, and with an insufficient appreciation of human factors. Many deficiencies may have arisen from a lack of competency in risk assessment or a lack of understanding of risk assessment by managers. Improvement in the use of risk assessment in the rail industry should be encouraged by providing guidance, adopting good practice risk assessment methods and training to improve understanding and competence in risk methods. The experts note Railtrack S&SD's development of the Safety Risk Model (SRM) and Controls Database. Of particular importance is the analysis of precursors and risk controls.

21. Future railway risk assessments should take more account of the complex interactions between the trains and the infrastructure. There should be better management and consideration of these complex, interactive interfaces. By way of clarification, all duty-holders must identify the hazards and risks within their control. This necessitates parties working together. There needs to be more sharing of information and greater transparency in the assessments.

22. Where possible high quality relevant local data on failure frequencies etc, should be used in risk assessments, supplemented, where necessary, by applicable generic data.

23. Although the problem of drawing meaningful conclusions from small sample sizes must be considered, the collection and analysis of local, company data should be encouraged and improved.

24. Benchmarking is important and should be encouraged as it assists in the sharing of information and best practice.

The ALARP principle, risk assessment and cost benefit analysis (CBA)

25. The "as low as reasonably practicable" (ALARP) principle is a sound concept for application in the rail industry. It should provide a powerful motivation for companies to seek continuous improvement. By way of explanation, the ALARP principle when used positively should assist companies in making the right safety related decisions. It should drive a company to consider what more can be done to further reduce the risk, i.e. just because some ambitious risk reduction scheme may not be reasonably practicable, that should not stop the investigation of less ambitious risk control schemes that are reasonably practicable. The benefits can be lost if a company attempts to prove solely that the cost of implementing a certain risk control is disproportionately high, without actively seeking any risk reduction. Some experts were concerned that the ALARP principle was being marginalised in the HSE discussion document, entitled "Reducing Risks, Protecting People"

26. The demonstration that risks are ALARP in current railway safety cases (RSCs) is lacking. Duty-holders should contend that risks are ALARP and explain the basis of their contention.

27. Detailed guidance on the demonstration that risks are ALARP and its application to the rail industry would be welcomed, because of the difficulty of demonstrating ALARP across the railway system.

28. CBA is an integral part of demonstrating that risks are ALARP. (It is noted that the subject of CBA was discussed extensively at the Joint Inquiry. This statement notes briefly the interaction of risk assessment with CBA). CBA is useful for examining a range of options for risk reduction. Risk assessment feeds into CBA by providing an estimate of the benefits. By way of explanation, risk assessment and CBA are inputs into the risk management decision-making process. CBA can be applied to consider a single option in isolation. It can also be used to compare options.

29. A CBA should clearly present its assumptions and findings so that it effectively communicates its results and uncertainties to decision-makers. It should be an explicit and transparent process.

30. More detailed guidance on CBA should be prepared for the rail industry.

Value and cost of preventing a fatality

31. To assess whether a risk control measure is reasonably practicable and that the expenditure is not in gross disproportion to the risk reduction, a trade off between costs and benefits is inescapable, either explicitly by the use of CBA with some "value of preventing a fatality" (VPF), or implicitly by use of "practices" that have proved cost-beneficial through experience or which themselves have been justified by CBA.

32. In assessing a range of risk control options, the practice in many industries is to divide the net cost of each option by the fatalities prevented. This gives a "cost of avoiding a fatality" for each option. These can be compared with a VPF. In this approach options can be ranked before defining a VPF.

33. An alternative approach is to calculate the value of safety benefits by multiplying the number of fatalities prevented by a VPF and comparing this with the net cost. This will give the same ranking as above.

34. For public expenditure and decisions in respect of public regulations the VPF should be based on evidence of the public's "willingness to pay".

35. A company may elect to use a higher VPF if it judges this to be in its commercial interests. Some experts observed that where a company is a monopoly or receives public money, in addition to having shareholder obligations and funding, then, a higher VPF, above the level of public preference, could not necessarily be justified. Other experts believed that a higher VPF, than one based on public "willingness to pay" alone, may be justified in such circumstances.

36. There is paradox in the application of VPF in the rail industry. By way of explanation, it is acknowledged that there are difficulties in determining a VPF. It is noted that studies of the kind undertaken by Professor Jones-Lee in this area indicate a VPF based on a public "willingness to pay" of about £1 million. In the experience of some experts, some companies (i.e. other industries) select to use a VPF of more than £10 million. The VPFs used in the rail industry are £1.15 million for an individual fatality and £3.22 million for multiple fatality events or where individual risks are close to intolerable. Several experts observed differences in government planned and contemplated expenditure e.g:

    (i)   the actual expenditure to prevent road fatality is around £0.1 million; and

    (ii)   the cost per fatality avoided by fitting TPWS is about £10 million.

If ATP were fitted after fitting TPWS the cost per fatality avoided is estimated to be at least £30 million.

37. Further research and debate is required into the question of whether society is, or is not, disproportionately averse to high consequence events. By way of explanation, this is an area of debate that affects all potentially hazardous activities. There is evidence (e.g. Professor Jones-Lee) that suggests that society is not disproportionately averse to high consequence events (catastrophic risks). There is no research evidence to the contrary. A fatality is a fatality. The application of weightings creates inequalities. A contrary view is held by some experts, who believe that it is appropriate to adopt higher VPFs to reflect what they believe is disproportionate social and political aversion to multiple fatality events.

Understanding risk assessment

38. There is a need to improve the general level of understanding of risk assessment in the rail industry. The knowledge and appreciation achieved should meet the needs of staff at all levels within an organisation. By way of clarification, safety-critical staff as well as supervisors and managers need to understand hazards and risks, and their control as an integral part of every task. The use of risk assessment can assist in gaining this understanding and knowledge. It is part of the process of communications that encourages ownership of the risks and drives towards continuous improvement.

Other issues

39. The annual safety case development plan, proposed in the Railways (Safety Case) Regulations 2000, should be incorporated into a company's annual safety plan rather than be a separate document. By way of clarification, a company's annual safety plan should set out their objectives, targets, priorities and actions planned for the forthcoming year. At present it includes how the company proposes to contribute to achieving the objectives in the annual Railway Group Safety Plan. To require a separate RSC Development Plan would be unwise as it should be an integral part of a company's overall plan, where there is one document that sets out the priorities for the year.

Description of risk assessment and risk management

1. As discussed in the document, the concept of risk is to be taken to include both the likelihood and consequences of undesired events, particularly those that would harm people.

2. Risk assessment is concerned with the identification of hazards and estimation of risks. The process includes:

    *   scoping – definition of the purpose of the risk assessment and scope of the issues to be addressed. This fundamental step is often overlooked or taken to be obvious;

    *   identification of hazards - what can go wrong? Systematic and comprehensive hazard identification is crucial for a robust risk assessment. A variety of methodologies exist (e.g. HAZOP – Hazard and Operability Study) to aid in this. Participation of all relevant disciplines is also important;

    *   assessment of likelihood or frequency - how often can it happen? Methods range from qualitative judgments to formal quantitative analysis;

    *   consequence assessment - how bad could the outcome be? Again, the assessments can be qualitative or quantitative. In safety risk assessments consequences are typically measured in terms of fatalities, major injuries, minor injuries, premature deaths, etc. Risk assessment consequences may be extended to other harms, such as business disruption and damage; and

    *   outputs from the risk assessment can be presented in a variety of ways - as a risk ranking matrix, as a risk log listing all identified hazards (with their assessed likelihood and consequences, etc), as a risk profile, as an ordered list of risks, in various diagrammatic forms, or as a series of fault-event trees.

3. Risk management addresses the appraisal of assessed risks and the making of decisions concerning risks, in particular safety measures and their subsequent implementation. It is a continual process, which includes:

    *   policy - aims and objectives to be achieved, informed by risk profile or other risk assessment outputs, and defining measures for success or failure;

    *   planning - the steps to be taken to manage the identified risks. Risks logs are often used as tools to record the risk controls and actions to be

applied to each identified hazard. Cost benefit analysis may be used to inform decisions on risk control;

- implementation - rolling out and executing the plan; and

- monitoring - checking on progress against objectives through audit, performance measurement, incident/accident investigation and management review.

Dissenting view

With reference to para 17, Mr O'Connor dissented and believes certain standards, such as ISO/IEC61508 (Functional Safety of Electrical/Electronic/Programmable Electronic Safety-related Systems) do not generally reflect current best practice and should not be adopted for application to the UK railway system without careful review and acceptance where appropriate by the regulators and the industry.

Appendix 8
Abbreviations

AAIB	Air Accidents Investigation Branch
ALARP	As Low As Reasonably Practicable
ASLEF	Associated Society of Locomotive Engineers and Firemen
ATOC	Association of Train Operating Companies
ATP	Automatic Train Protection
AWS	Automatic Warning System
BRIS	British Rail Infrastructure Services
BTP	British Transport Police
CAA	Civil Aviation Authority
CIRAS	Confidential Incident Reporting and Analysis System
DETR	Department of the Environment, Transport and the Regions *(now DTLR)*
DOO	Driver-Only Operation
DTLR	Department of Transport, Local Government and the Regions *(formerly DETR)*
ERTMS	European Rail Traffic Management System
ETSC	European Transport Safety Council
EWS	English, Welsh and Scottish Railway
FAA	Federal Aviation Administration
FGW	First Great Western
FOC	Freight Train Operating Company
GNER	Great North Eastern Railway
HMRI	Her Majesty's Railway Inspectorate
HSC	Health and Safety Commission
HSE	Health and Safety Executive
HST	High Speed Train
IMC	Infrastructure Maintenance Company
IOSH	Institution of Occupational Safety and Health
IRO	Institution of Railway Operators
IRSA	Independent Railway Safety Activity
IRSE	Institution of Railway Signal Engineers
LUL	London Underground Ltd
MAIB	Marine Accident Investigation Branch
MOLA	Master Operating Lease Agreement
NERA	National Economic Research Associates
NRSA	National Rail Safety Authority
NTSB	National Transportation Safety Board
NVQ	National Vocational Qualification
OPRAF	Office of Passenger Rail Franchising
ORR	Office of the Rail Regulator
RAIB	Railway Accident Investigation Branch
RI	Railway Inspectorate
RIA	Railway Industry Association
RIAC	Rail Industry Advisory Committee
RISB	Rail Industry Safety Body

RISSC	Railway Industry Standards Strategy Committee
RITC	Rail Industry Training Council
RMT	National Union of Rail, Maritime and Transport Workers
ROSCO	Rolling Stock Company
RSC	Railway Safety Case
S&SD	Safety and Standards Directorate
SAB	Safety Advisory Board
SMIS	Safety Management Information System
SPAD	Signal Passed at Danger
SRA	Strategic Rail Authority
SRG	Safety Regulation Group
TESCO	Technical Support Company
TOC	*(passenger)* Train Operating Company
TPWS	Train Protection Warning System
TQM	Total Quality Management
TRC	Track Renewal Company
TSR	Transport Safety Review *(team of the DETR)*
TSSA	Transport Salaried Staffs' Association

Inquiry team

Chairman

The Rt Hon Lord Cullen PC

Assessors

Professor Peter McKie CBE
Malcolm Southgate

Counsel to the Inquiry

Robert Owen QC
Neil Garnham
Eric Brown
Susan Chan

assisted by

Rebecca Cope

Internal consultant on safety matters

Rod Sylvester-Evans

Solicitors

Michael Fitzgerald
Myles Hothersall

Secretariat

Andrew Allberry
Craig Frost
Ruth Gohler
Annie Johnston
David Steer
Sunil Wickramaratne

Josephine Fowles
Monica Garcia
Simon Harris
Peter O'Connor
Jansen Versfeld

The Inquiry team was grateful for the technical advice and assistance of Major C B Holden OBE.